An Introduction to Mathematical Finance

This elementary introduction to the theory of options pricing presents the Black–Scholes theory of options as well as such general topics in finance as the time value of money, rate of return of an investment cash-flow sequence, utility functions and expected utility maximization, mean variance analysis, optimal portfolio selection, and the capital assets pricing model.

The author assumes no prior knowledge of probability and presents all the necessary preliminary material simply and clearly in chapters on probability, normal random variables, and the geometric Brownian motion model that underlies the Black–Scholes theory. He carefully explains the concept of arbitrage, using many examples, and he then presents the arbitrage theorem and uses it, along with a multiperiod binomial approximation of geometric Brownian motion, to obtain a simple derivation of the Black–Scholes call option formula. Later chapters treat risk-neutral (nonarbitrage) pricing of exotic options – both by Monte Carlo simulation and by multiperiod binomial approximation models for European and American style options. Finally, the author presents real price data indicating that the underlying geometric Brownian motion model is not always appropriate and shows how the model can be generalized to deal with such situations.

No other text presents such sophisticated topics in a mathematically accurate but accessible way. This book will appeal to professional traders as well as to undergraduates studying the basics of finance.

Sheldon M. Ross is a professor in the Department of Industrial Engineering and Operations Research at the University of California at Berkeley. He received his Ph.D. in statistics at Stanford University in 1968 and has been at Berkeley ever since. He has published nearly 100 articles and a variety of textbooks in the areas of statistics and applied probability. He is the founding and continuing editor of the journal *Probability in the Engineering and Informational Sciences,* a fellow of the Institute of Mathematical Statistics, and a recipient of the Humboldt U.S. Senior Scientist Award.

An Introduction to Mathematical Finance

Options and Other Topics

SHELDON M. ROSS
University of California at Berkeley

CAMBRIDGE
UNIVERSITY PRESS

PUBLISHED BY THE PRESS SYNDICATE OF THE UNIVERSITY OF CAMBRIDGE
The Pitt Building, Trumpington Street, Cambridge, United Kingdom

CAMBRIDGE UNIVERSITY PRESS
The Edinburgh Building, Cambridge CB2 2RU, UK   www.cup.cam.ac.uk
40 West 20th Street, New York, NY 10011-4211, USA   www.cup.org
10 Stamford Road, Oakleigh, Melbourne 3166, Australia
Ruiz de Alarcón 13, 28014 Madrid, Spain

© Cambridge University Press 1999

First published 1999

Printed in the United States of America

Typeface Times 11/14 pt.   *System* AMS-T$_E$X   [FH]

A catalog record for this book is available from the British Library

Library of Congress Cataloging in Publication Data
Ross, Sheldon M.
An introduction to mathematical finance : options and other topics
/ Sheldon M. Ross.
p.   cm.
ISBN 0-521-77043-2 (hardbound)
1. Investments – Mathematics.   2. Stochastic analysis.   3. Options
(Finance) – Mathematical models.   4. Securities prices – Mathematical
models.   I. Title.
HG4515.8.R67   1999
332.6'01'61 – dc21   99-25389
                                                              CIP

ISBN 0 521 77043 2 hardback

To my parents,
Ethel and Louis Ross

Contents

Introduction and Preface

An *option* gives one the right, but not the obligation, to buy or sell a security under specified terms. A *call option* is one that gives the right to buy, and a *put option* is one that gives the right to sell the security. Both types of options will have an *exercise price* and an *exercise time*. In addition, there are two standard conditions under which options operate: *European* options can be utilized only at the exercise time, whereas *American* options can be utilized at any time up to the exercise time. Thus, for instance, a European call option with exercise price K and exercise time t gives its holder the right to purchase at time t one share of the underlying security for the price K, whereas an American call option gives its holder the right to make that purchase at any time before or at time t.

A prerequisite for a strong market in options is a computationally efficient way of evaluating, at least approximately, their worth; this was accomplished for call options (of either American or European type) by the famous Black–Scholes formula. The formula assumes that prices of the underlying security follow a geometric Brownian motion. This means that if $S(y)$ is the price of the security at time y then, for any price history up to time y, the ratio of the price at a specified future time $t + y$ to the price at time y has a lognormal distribution with mean and variance parameters $t\mu$ and $t\sigma^2$, respectively. That is,

$$\log\left(\frac{S(t + y)}{S(y)}\right)$$

will be a normal random variable with mean $t\mu$ and variance $t\sigma^2$. Black and Scholes showed, under the assumption that the prices follow a geometric Brownian motion, that there is a single price for a call option that does not allow an idealized trader – one who can instantaneously make trades without any transaction costs – to follow a strategy that will result in a sure profit in all cases. That is, there will be no certain profit (i.e., no *arbitrage*) if and only if the price of the option is as given by the Black–Scholes formula. In addition, this price depends only on the variance parameter σ of the geometric Brownian motion (as well as on the prevailing interest rate, the underlying price of the security, and the conditions of the option) and not on the parameter μ. Because the parameter σ is a measure of the volatility of the security, it is often called the *volatility* parameter.

A *risk-neutral* investor is one who values an investment solely through the expected present value of its return. If such an investor models a security by a geometric Brownian motion that turns all investments involving buying and selling the security into fair bets, then this investor's valuation of a call option on this security will be precisely as given by the Black–Scholes formula. For this reason, the Black–Scholes valuation is often called a *risk-neutral valuation*.

Our first objective in this book is to derive and explain the Black–Scholes formula. This does require some knowledge of probability, the topic considered in the first three chapters. Chapter 1 introduces probability and the probability experiment. Random variables – numerical quantities whose values are determined by the outcome of the probability experiment – are discussed, as are the concepts of the expected value and variance of a random variable. In Chapter 2 we introduce normal random variables; these are random variables whose probabilities are determined by a bell-shaped curve. The central limit theorem is presented in this chapter. This theorem, probably the most important theoretical result in probability, states that the sum of a large number of random variables will approximately be a normal random variable. In Chapter 3 we introduce the geometric Brownian motion process; we define it, show how it can be obtained as the limit of simpler processes, and discuss the justification for its use in modeling security prices.

With the probability necessities behind us, the second part of the text begins in Chapter 4 with an introduction to the concept of interest rates and present values. A key concept underlying the Black–Scholes formula is that of arbitrage, the subject of Chapter 5. In this chapter we show how arbitrage can be used to determine prices in a variety of situations, including the single-period binomial option model. In Chapter 6 we present the arbitrage theorem and use it to find an expression for the unique nonarbitrage option cost in the multiperiod binomial model. In Chapter 7 we use the results of Chapter 6, along with the approximations of geometric Brownian motion presented in Chapter 3, to obtain a simple derivation of the Black–Scholes equation for pricing call options. In addition, we show how to utilize a multiperiod binomial model to determine an approximation of the risk-neutral price of an American put option.

In Chapter 8 we note that, in many situations, arbitrage considerations do not result in a unique cost. In such cases we show the importance of the investor's utility function as well as his or her estimates of the probabilities of the possible outcomes of the investment. Applications are given to portfolio selection problems, and the capital assets pricing model is introduced. In addition we show that, even when a security's price follows a geometric Brownian motion and call options are priced according to the Black–Scholes formula, there may still be investment opportunities that have a positive expected gain with a relatively small standard deviation. (Such opportunities arise when an investor's evaluation of the geometric Brownian motion parameter μ differs from the value that turns all investment bets into fair bets.)

In Chapter 9 we introduce some nonstandard, or "exotic," options such as barrier, Asian, and lookback options. We explain how to use Monte Carlo simulation techniques to efficiently determine the geometric Brownian motion risk-neutral valuation of such options. Our ways of exploiting variance reduction ideas to make the simulation more efficient have not previously appeared and are improvements over what is presently in the literature.

The Black–Scholes formula is useful even if one has doubts about the validity of the underlying geometric Brownian model. For as long as one accepts that this model is at least approximately valid, its use suggests the *appropriate* price of the option. Thus, if the actual trading option price is below the formula price then it would seem that the option is underpriced in relation to the security itself, thus leading one to consider a strategy of buying options and selling the security (with the reverse being suggested when the trading option price is above the formula price). However, one downside to the Black–Scholes formula is that its very usefulness and computational simplicity has led many to automatically assume the underlying geometric Brownian motion model; as a result, relatively little effort has gone into searching for a better model. In Chapter 10 we show that real data cannot aways be fit by a geometric Brownian motion model, and that more general models may need to be considered. For instance, one of the key assumptions of geo-metric Brownian motion is that the ratio of a future security price to the present price does not depend on past prices. In Chapter 10, we consider approximately 3 years of data concerning the (nearest-month) price of crude oil. Each day is characterized as being of one of four types: type 1 means that today's final crude price is down from yesterday's by more than 1%; type 2 means that the price is down by less than 1%; type 3 means that it is up by less than 1%; and type 4 that it is up by more than 1%. The following table gives the percentage of time that a type-i day was followed by a type-j day for $i, j = 1, \ldots, 4$.

		j		
i	1	2	3	4
1	31	23	25	21
2	21	30	21	28
3	15	28	28	29
4	27	32	16	25

Thus, for instance, a large drop (greater than 1%) was followed 31% of the time by another large drop, 23% of the time by a small drop, 25% of the time by a small increase, and 21% percent of the time by a large increase. Under the geometric Brownian motion model, tomor-row's change would be unaffected by today's, and so the theoretically expected percentages in the preceding table would be the same for all rows. A standard statistical procedure indicates that, if the row probabil-ities were equal (as implied by geometric Brownian motion), then data as nonsupportive of this hypothesized equality as the data actually ob-tained would occur only .5% of the time. Consequently, the hypothesis that the prices of crude oil follow geometric Brownian motion is re-jected. In Chapter 10 we then formulate an improved model that is both intuitively reasonable and (most importantly) fits the data better than geometric Brownian motion, and we show how to obtain a risk-neutral option valuation based on this improved model.

In the case of commodity prices, there is a strong belief by many traders in the concept of mean price reversion: that the market prices of certain commodities have tendencies to revert to fixed values. In

Chapter 11 we present a model, more general than geometric Brownian motion, that can be used to model the price flow of such a commodity.

One technical point that should be mentioned is that we use the notation $\log(x)$ to represent the natural logarithm of x. That is, the logarithm has base e, where e is defined by

$$e = \lim_{n \to \infty} (1 + 1/n)^n$$

and is approximately given by $2.71828\ldots$.

We would like to thank Professors Ilan Adler and Shmuel Oren for some enlightening conversations, Mr. Kyle Lin for his many useful comments, and Mr. Nahoya Takezawa for his general comments and for doing the numerical work needed in the final chapters.

1. Probability

1.1 Probabilities and Events

Consider an experiment and let S, called the *sample space,* be the set of all possible outcomes of the experiment. If there are m possible outcomes of the experiment then we will generally number them 1 through m, and so $S = \{1, 2, \ldots, m\}$. However, when dealing with specific examples, we will usually give more descriptive names to the outcomes.

Example 1.1a (i) Let the experiment consist of flipping a coin, and let the outcome be the side that lands face up. Thus, the sample space of this experiment is

$$S = \{h, t\},$$

where the outcome is h if the coin shows heads and t if it shows tails.

(ii) If the experiment consists of rolling a pair of dice – with the outcome being the pair (i, j), where i is the value that appears on the first die and j the value on the second – then the sample space consists of the following 36 outcomes:

$$(1, 1), \ (1, 2), \ (1, 3), \ (1, 4), \ (1, 5), \ (1, 6),$$
$$(2, 1), \ (2, 2), \ (2, 3), \ (2, 4), \ (2, 5), \ (2, 6),$$
$$(3, 1), \ (3, 2), \ (3, 3), \ (3, 4), \ (3, 5), \ (3, 6),$$
$$(4, 1), \ (4, 2), \ (4, 3), \ (4, 4), \ (4, 5), \ (4, 6),$$
$$(5, 1), \ (5, 2), \ (5, 3), \ (5, 4), \ (5, 5), \ (5, 6),$$
$$(6, 1), \ (6, 2), \ (6, 3), \ (6, 4), \ (6, 5), \ (6, 6).$$

(iii) If the experiment consists of a race of r horses numbered 1, 2, 3, $\ldots, r$, and the outcome is the order of finish of these horses, then the sample space is

$$S = \{\text{all orderings of the numbers } 1, 2, 3, \ldots, r\}.$$

For instance, if $r = 4$ then the outcome is $(1, 4, 2, 3)$ if the number-1 horse comes in first, number 4 comes in second, number 2 comes in third, and number 3 comes in fourth.   □

Consider once again an experiment with the sample space $S = \{1, 2, \ldots, m\}$. We will now suppose that there are numbers $p_1, \ldots, p_m$ with

$$p_i \geq 0, \quad i = 1, \ldots, m, \quad \text{and} \quad \sum_{i=1}^{m} p_i = 1$$

and such that p_i is the *probability* that i is the outcome of the experiment.

Example 1.1b   In Example 1.1a(i), the coin is said to be *fair* or *unbiased* if it is equally likely to land on heads as on tails. Thus, for a fair coin we would have that

$$p_h = p_t = 1/2.$$

If the coin were biased and heads were twice as likely to appear as tails, then we would have

$$p_h = 2/3, \qquad p_t = 1/3.$$

If an unbiased pair of dice were rolled in Example 1.1a(ii), then all possible outcomes would be equally likely and so

$$p_{(i,j)} = 1/36, \quad 1 \leq i \leq 6, \ 1 \leq j \leq 6.$$

If $r = 3$ in Example 1.1a(iii), then we suppose that we are given the six nonnegative numbers that sum to 1:

$$p_{1,2,3}, \ p_{1,3,2}, \ p_{2,1,3}, \ p_{2,3,1}, \ p_{3,1,2}, \ p_{3,2,1},$$

where $p_{i,j,k}$ represents the probability that horse i comes in first, horse j second, and horse k third.   □

Any set of possible outcomes of the experiment is called an *event*. That is, an event is a subset of S, the set of all possible outcomes. For any event A, we say that A *occurs* whenever the outcome of the experiment is a point in A. If we let $P(A)$ denote the probability that event A occurs, then we can determine it by using the equation

$$P(A) = \sum_{i \in A} p_i. \tag{1.1}$$

Note that this implies

$$P(S) = \sum_i p_i = 1. \tag{1.2}$$

In words, the probability that the outcome of the experiment is in the sample space is equal to 1 – which, since S consists of all possible outcomes of the experiment, is the desired result.

Example 1.1c   Suppose the experiment consists of rolling a pair of fair dice. If A is the event that the sum of the dice is equal to 7, then

$$A = \{(1, 6), (2, 5), (3, 4), (4, 3), (5, 2), (6, 1)\}$$

and

$$P(A) = 6/36 = 1/6.$$

If we let B be the event that the sum is 8, then

$$P(B) = p_{(2,6)} + p_{(3,5)} + p_{(4,4)} + p_{(5,3)} + p_{(6,2)} = 5/36.$$

If, in a horse race between three horses, we let A denote the event that horse number 1 wins, then $A = \{(1, 2, 3), (1, 3, 2)\}$ and

$$P(A) = p_{1,2,3} + p_{1,3,2}. \qquad \square$$

For any event A, we let A^c, called the *complement* of A, be the event containing all those outcomes in S that are not in A. That is, A^c occurs if and only if A does not. Since

$$1 = \sum_i p_i$$
$$= \sum_{i \in A} p_i + \sum_{i \in A^c} p_i$$
$$= P(A) + P(A^c),$$

we see that

$$P(A^c) = 1 - P(A). \tag{1.3}$$

That is, the probability that the outcome is not in A is 1 minus the probability that it is in A. The complement of the sample space S is the null event $\emptyset$, which contains no outcomes. Since $\emptyset = S^c$, we obtain from Equations (1.2) and (1.3) that

$$P(\emptyset) = 0.$$

For any events A and B we define $A \cup B$, called the *union* of A and B, as the event consisting of all outcomes that are in A, or in B, or in both A and B. Also, we define their *intersection* AB (sometimes written $A \cap B$) as the event consisting of all outcomes that are both in A and in B.

Example 1.1d   Let the experiment consist of rolling a pair of dice. If A is the event that the sum is 10 and B is the event that both dice land on even numbers greater than 3, then

$$A = \{(4, 6), (5, 5), (6, 4)\}, \qquad B = \{(4, 4), (4, 6), (6, 4), (6, 6)\}.$$

Therefore,

$$A \cup B = \{(4, 4), (4, 6), (5, 5), (6, 4), (6, 6)\},$$

$$AB = \{(4, 6), (6, 4)\}. \qquad \square$$

For any events A and B, we can write

$$P(A \cup B) = \sum_{i \in A \cup B} p_i,$$

$$P(A) = \sum_{i \in A} p_i,$$

$$P(B) = \sum_{i \in B} p_i.$$

Since every outcome in both A and B is counted twice in $P(A) + P(B)$ and only once in $P(A \cup B)$, we obtain the following result, often called the *addition theorem of probability*.

Proposition 1.1.1

$$P(A \cup B) = P(A) + P(B) - P(AB).$$

Thus, the probability that the outcome of the experiment is either in A or in B is: the probability that it is in A, plus the probability that it is in B, minus the probability that it is in both A and B.

Example 1.1e Suppose the probabilities that the Dow-Jones stock index increases today is .54, that it increases tomorrow is .54, and that it increases both days is .28. What is the probability that it does not increase on either day?

Solution. Let A be the event that the index increases today, and let B be the event that it increases tomorrow. Then the probability that it increases on at least one of these days is

$$P(A \cup B) = P(A) + P(B) - P(AB)$$

$$= .54 + .54 - .28 = .80.$$

Therefore, the probability that it increases on neither day is $1 - .80 = .20$. $\qquad\square$

If $AB = \emptyset$, we say that A and B are *mutually exclusive* or *disjoint*. That is, events are mutually exclusive if they cannot both occur. Since $P(\emptyset) = 0$, it follows from Proposition 1.1.1 that, when A and B are mutually exclusive,

$$P(A \cup B) = P(A) + P(B).$$

1.2   Conditional Probability

Suppose that each of two teams is to produce an item, and that the two items produced will be rated as either acceptable or unacceptable. The sample space of this experiment will then consist of the following four outcomes:

$$S = \{(a, a), (a, u), (u, a), (u, u)\},$$

where (a, u) means, for instance, that the first team produced an acceptable item and the second team an unacceptable one. Suppose that the probabilities of these outcomes are as follows:

$$P(a, a) = .54,$$

$$P(a, u) = .28,$$

$$P(u, a) = .14,$$

$$P(u, u) = .04.$$

If we are given the information that exactly one of the items produced was acceptable, what is the probability that it was the one produced by the first team? To determine this probability, consider the following reasoning. Given that there was exactly one acceptable item produced, it follows that the outcome of the experiment was either (a, u) or (u, a). Since the outcome (a, u) was initially twice as likely as the outcome (u, a), it should remain twice as likely given the information that one of them occurred. Therefore, the probability that the outcome was (a, u) is $2/3$, whereas the probability that it was (u, a) is $1/3$.

Let $A = \{(a, u), (a, a)\}$ denote the event that the item produced by the first team is acceptable, and let $B = \{(a, u), (u, a)\}$ be the event that exactly one of the produced items is acceptable. The probability that the item produced by the first team was acceptable given that exactly one of the produced items was acceptable is called the *conditional probability* of A given that B has occurred; this is denoted as

$$P(A|B).$$

A general formula for $P(A|B)$ is obtained by an argument similar to the one given in the preceding. Namely, if the event B occurs then, in order for the event A to occur, it is necessary that the occurrence be a point in both A and B; that is, it must be in AB. Now, since we know that B has occurred, it follows that B can be thought of as the new sample space, and hence the probability that the event AB occurs will equal the probability of AB relative to the probability of B. That is,

$$P(A|B) = \frac{P(AB)}{P(B)}. \tag{1.4}$$

Example 1.2a   A coin is flipped twice. Assuming that all four points in the sample space $S = \{(h, h), (h, t), (t, h), (t, t)\}$ are equally likely, what is the conditional probability that both flips land on heads, given that

(a) the first flip lands on heads, and
(b) at least one of the flips lands on heads?

Solution. Let $A = \{(h, h)\}$ be the event that both flips land on heads; let $B = \{(h, h), (h, t)\}$ be the event that the first flip lands on heads; and let $C = \{(h, h), (h, t), (t, h)\}$ be the event that at least one of the flips lands on heads. We have the following solutions:

$$P(A|B) = \frac{P(AB)}{P(B)}$$

$$= \frac{P(\{(h, h)\})}{P(\{(h, h), (h, t)\})}$$

$$= \frac{1/4}{2/4}$$

$$= 1/2$$

and

$$P(A|C) = \frac{P(AC)}{P(C)}$$

$$= \frac{P(\{(h, h)\})}{P(\{(h, h), (h, t), (t, h)\})}$$

$$= \frac{1/4}{3/4}$$

$$= 1/3.$$

Many people are initially surprised that the answers to parts (a) and (b) are not identical. To understand why the answers are different, note first that – conditional on the first flip landing on heads – the second one is still equally likely to land on either heads or tails, and so the probability in part (a) is $1/2$. On the other hand, knowing that at least one of the flips lands on heads is equivalent to knowing that the outcome is not (t, t). Thus, given that at least one of the flips lands on heads, there remain three equally likely possibilities, namely (h, h), (h, t), (t, h), showing that the answer to part (b) is $1/3$. □

It follows from Equation (1.4) that

$$P(AB) = P(B)P(A|B). \tag{1.5}$$

That is, the probability that both A and B occur is the probability that B occurs multiplied by the conditional probability that A occurs given that B occurred; this result is often called the *multiplication theorem of probability*.

Example 1.2b   Suppose that two balls are to be withdrawn, without replacement, from an urn that contains 9 blue and 7 yellow balls. If each ball drawn is equally likely to be any of the balls in the urn at the time, what is the probability that both balls are blue?

Solution. Let B_1 and B_2 denote, respectively, the events that the first and second balls withdrawn are blue. Now, given that the first ball withdrawn is blue, the second ball is equally likely to be any of the remaining 15 balls, of which 8 are blue. Therefore, $P(B_2|B_1) = 8/15$. As $P(B_1) = 9/16$, we see that

$$P(B_1B_2) = \frac{9}{16}\frac{8}{15} = \frac{3}{10}.$$ □

The conditional probability of A given that B has occurred is not generally equal to the unconditional probability of A. In other words, knowing that the outcome of the experment is an element of B generally changes the probability that it is an element of A. (What if A and B are mutually exclusive?) In the special case where $P(A|B)$ is equal to $P(A)$, we say that A is *independent* of B. Since

$$P(A|B) = \frac{P(AB)}{P(B)},$$

we see that A is independent of B if

$$P(AB) = P(A)P(B). \tag{1.6}$$

The relation in (1.6) is symmetric in A and B. Thus it follows that, whenever A is independent of B, B is also independent of A – that is, A and B are *independent events*.

Example 1.2c   Suppose that, with probability .52, the closing price of a stock is at least as high as the close on the previous day, and that the results for succesive days are independent. Find the probability that the closing price goes down in each of the next four days, but not on the following day.

Solution. Let A_i be the event that the closing price goes down on day i. Then, by independence, we have

$$P(A_1A_2A_3A_4A_5^c) = P(A_1)P(A_2)P(A_3)P(A_4)P(A_5^c)$$

$$= (.48)^4(.52) = .0276.$$ □

1.3   Random Variables and Expected Values

Numerical quantities whose values are determined by the outcome of the experiment are known as *random variables*. For instance, the sum obtained when rolling dice, or the number of heads that result in a series of coin flips, are random variables. Since the value of a random variable is determined by the outcome of the experiment, we can assign probabilities to each of its possible values.

Example 1.3a   Let the random variable X denote the sum when a pair of fair dice are rolled. The possible values of X are $2, 3, \ldots, 12$, and they have the following probabilities:

$P\{X = 2\} = P\{(1, 1)\} = 1/36,$

$P\{X = 3\} = P\{(1, 2), (2, 1)\} = 2/36,$

$P\{X = 4\} = P\{(1, 3), (2, 2), (3, 1)\} = 3/36,$

$P\{X = 5\} = P\{(1, 4), (2, 3), (3, 2), (4, 1)\} = 4/36,$

$P\{X = 6\} = P\{(1, 5), (2, 4), (3, 3), (4, 2), (5, 1)\} = 5/36,$

$P\{X = 7\} = P\{(1, 6), (2, 5), (3, 4), (4, 3), (5, 2), (6, 1)\} = 6/36,$

$P\{X = 8\} = P\{(2, 6), (3, 5), (4, 4), (5, 3), (6, 2)\} = 5/36,$

$P\{X = 9\} = P\{(3, 6), (4, 5), (5, 4), (6, 3)\} = 4/36,$

$P\{X = 10\} = P\{(4, 6), (5, 5), (6, 4)\} = 3/36,$

$P\{X = 11\} = P\{(5, 6), (6, 5)\} = 2/36,$

$P\{X = 12\} = P\{(6, 6)\} = 1/36.$ □

If X is a random variable whose possible values are $x_1, x_2, \ldots, x_n$, then the set of probabilities $P\{X = x_j\}$ $(j = 1, \ldots, n)$ is called the *probability distribution* of the random variable. Since X must assume one of these values, it follows that

$$\sum_{j=1}^{n} P\{X = x_j\} = 1.$$

Definition   If X is a random variable whose possible values are $x_1, x_2, \ldots, x_n$, then the *expected value* of X, denoted by $E[X]$, is defined by

$$E[X] = \sum_{j=1}^{n} x_j P\{X = x_j\}.$$

Alternative names for $E[X]$ are the *expectation* or the *mean* of X.

In words, $E[X]$ is a weighted average of the possible values of X, where the weight given to a value is equal to the probability that X assumes that value.

Example 1.3b   Let the random variable X denote the amount that we win when we make a certain bet. Find $E[X]$ if there is a 60% chance that we lose 1, a 20% chance that we win 1, and a 20% chance that we win 2.

Solution.
$$E[X] = -1(.6) + 1(.2) + 2(.2) = 0.$$

Thus, the expected amount that is won on this bet is equal to 0. A bet whose expected winnings is equal to 0 is called a *fair* bet.   □

Example 1.3c   A random variable X, which is equal to 1 with probability p and to 0 with probability $1 - p$, is said to be a *Bernoulli* random variable with parameter p. Its expected value is

$$E[X] = 1(p) + 0(1 - p) = p.$$ □

A useful and easily established result is that, for constants a and b,

$$E[aX + b] = aE[X] + b. \qquad (1.7)$$

To verify Equation (1.7), let $Y = aX + b$. Since Y will equal $ax_j + b$ when $X = x_j$, it follows that

$$
\begin{aligned}
E[Y] &= \sum_{j=1}^{n} (ax_j + b) P\{X = x_j\} \\
&= \sum_{j=1}^{n} ax_j P\{X = x_j\} + \sum_{j=1}^{n} bP\{X = x_j\} \\
&= a \sum_{j=1}^{n} x_j P\{X = x_j\} + b \sum_{j=1}^{n} P\{X = x_j\} \\
&= aE[X] + b.
\end{aligned}
$$

An important result is that the expected value of a sum of random variables is equal to the sum of their expected values.

Proposition 1.3.1   *For random variables $X_1, \ldots, X_k$,*

$$E\left[\sum_{j=1}^{k} X_j\right] = \sum_{j=1}^{k} E[X_j].$$

Example 1.3d   Consider n independent trials, each of which is a success with probability p. The random variable X, equal to the total number of successes that occur, is called a *binomial* random variable with parameters n and p. We can determine its expectation by using the representation

$$X = \sum_{j=1}^{n} X_j,$$

where X_j is defined to equal 1 if trial j is a success and to equal 0 otherwise. Using Proposition 1.3.1, we obtain that

$$E[X] = \sum_{j=1}^{n} E[X_j] = np,$$

where the final equality used the result of Example 1.3c.   □

The random variables $X_1, \ldots, X_n$ are said to be *independent* if probabilities concerning any subset of them are unchanged by information as to the values of the others.

Example 1.3e   Suppose that k balls are to be randomly chosen from a set of N balls, of which n are red. If we let X_i equal 1 if the ith ball chosen is red and 0 if it is black, then $X_1, \ldots, X_n$ would be independent if each selected ball is replaced before the next selection is made but they would not be independent if each selection is made without replacing previously selected balls. (Why not?)   □

Whereas the average of the possible values of X is indicated by its expected value, its spread is measured by its variance.

Definition   The *variance* of X, denoted by $\text{Var}(X)$, is defined by

$$\text{Var}(X) = E[(X - E[X])^2].$$

In other words, the variance measures the average square of the difference between X and its expected value.

Example 1.3f   Find $\text{Var}(X)$ when X is a Bernoulli random variable with parameter p.

Solution.   Because $E[X] = p$ (as shown in Example 1.3c), we see that

$$(X - E[X])^2 = \begin{cases} (1-p)^2 & \text{with probability } p \\ p^2 & \text{with probability } 1-p. \end{cases}$$

Hence,

$$\begin{aligned} \text{Var}(X) &= E[(X - E[X])^2] \\ &= (1-p)^2 p + p^2(1-p) \\ &= p - p^2. \end{aligned} \qquad \Box$$

If a and b are constants, then

$$\begin{aligned} \text{Var}(aX + b) &= E[(aX + b - E[aX + b])^2] \\ &= E[(aX - aE[X])^2] \qquad \text{(by Equation (1.7))} \\ &= E[a^2(X - E[X])^2] \\ &= a^2 \text{Var}(X). \end{aligned} \qquad (1.8)$$

Although it is not generally true that the variance of the sum of random variables is equal to the sum of their variances, this *is* the case when the random variables are independent.

Proposition 1.3.2   *If $X_1, \ldots, X_k$ are independent random variables, then*

$$\text{Var}\left(\sum_{j=1}^{k} X_j\right) = \sum_{j=1}^{k} \text{Var}(X_j).$$

Example 1.3g   Find the variance of X, a binomial random variable with parameters n and p.

Solution. Recalling that X represents the number of successes in n *independent* trials (each of which is a success with probability p), we can represent it as

$$X = \sum_{j=1}^{n} X_j,$$

where X_j is defined to equal 1 if trial j is a success and 0 otherwise. Hence,

$$\text{Var}(X) = \sum_{j=1}^{n} \text{Var}(X_j) \qquad \text{(by Proposition 1.3.2)}$$

$$= \sum_{j=1}^{n} p(1-p) \qquad \text{(by Example 1.3f)}$$

$$= np(1-p). \qquad \qquad \square$$

The square root of the variance is called the *standard deviation*. As we shall see, a random variable tends to lie within a few standard deviations of its expected value.

1.4   Covariance and Correlation

The covariance of any two random variables X and Y, denoted by $\text{Cov}(X, Y)$, is defined by

$$\text{Cov}(X, Y) = E[(X - E[X])(Y - E[Y])].$$

Upon multiplying the terms within the expectation, and then taking expectation term by term, it can be shown that

$$\text{Cov}(X, Y) = E[XY] - E[X]E[Y].$$

A positive value of the covariance indicates that X and Y both tend to be large at the same time, whereas a negative value indicates that when one is large the other tends to be small. (Independent random variables have covariance equal to 0.)

Example 1.4a   Let X and Y both be Bernoulli random variables. That is, each takes on either the value 0 or 1. Using the identity

$$\text{Cov}(X, Y) = E[XY] - E[X]E[Y]$$

and noting that XY will equal 1 or 0 depending upon whether both X and Y are equal to 1, we obtain that

$$\text{Cov}(X, Y) = P\{X = 1, \ Y = 1\} - P\{X = 1\}P\{Y = 1\}.$$

From this, we see that

$$\text{Cov}(X, Y) > 0 \iff P\{X = 1, \ Y = 1\} > P\{X = 1\}P\{Y = 1\}$$

$$\iff \frac{P\{X = 1, \ Y = 1\}}{P\{X = 1\}} > P\{Y = 1\}$$

$$\iff P\{Y = 1 \mid X = 1\} > P\{Y = 1\}.$$

That is, the covariance of X and Y is positive if the outcome that $X = 1$ makes it more likely that $Y = 1$ (which, as is easily seen, also implies the reverse). □

The following properties of covariance are easily established. For random variables X and Y, and constant c:

$$\text{Cov}(X, Y) = \text{Cov}(Y, X),$$

$$\text{Cov}(X, X) = \text{Var}(X),$$

$$\text{Cov}(cX, Y) = c\,\text{Cov}(X, Y),$$

$$\text{Cov}(c, Y) = 0.$$

Covariance, like expected value, satisfies a linearity property – namely,

$$\text{Cov}(X_1 + X_2, Y) = \text{Cov}(X_1, Y) + \text{Cov}(X_2, Y). \qquad (1.9)$$

Equation (1.9) is proven as follows:

$$\begin{aligned}
\text{Cov}(X_1 + X_2, Y) &= E[(X_1 + X_2)Y] - E[X_1 + X_2]E[Y] \\
&= E[X_1Y + X_2Y] - (E[X_1] + E[X_2])E[Y] \\
&= E[X_1Y] - E[X_1]E[Y] + E[X_2Y] - E[X_2]E[Y] \\
&= \text{Cov}(X_1, Y) + \text{Cov}(X_2, Y).
\end{aligned}$$

Equation (1.9) is easily generalized to yield the following useful identity:

$$\text{Cov}\left(\sum_{i=1}^{n} X_i, \sum_{j=1}^{m} Y_j\right) = \sum_{i=1}^{n} \sum_{j=1}^{m} \text{Cov}(X_i, Y_j). \qquad (1.10)$$

Equation (1.10) yields a useful formula for the variance of the sum of random variables:

$$\text{Var}\left(\sum_{i=1}^{n} X_i\right) = \text{Cov}\left(\sum_{i=1}^{n} X_i, \sum_{j=1}^{n} X_j\right)$$

$$= \sum_{i=1}^{n} \sum_{j=1}^{n} \text{Cov}(X_i, X_j)$$

$$= \sum_{i=1}^{n} \text{Cov}(X_i, X_i) + \sum_{i=1}^{n} \sum_{j \neq i} \text{Cov}(X_i, X_j)$$

$$= \sum_{i=1}^{n} \text{Var}(X_i) + \sum_{i=1}^{n} \sum_{j \neq i} \text{Cov}(X_i, X_j). \qquad (1.11)$$

The degree to which large values of X tend to be associated with large values of Y is measured by the *correlation* between X and Y, denoted as $\rho(X, Y)$ and defined by

$$\rho(X, Y) = \frac{\text{Cov}(X, Y)}{\sqrt{\text{Var}(X)\,\text{Var}(Y)}}.$$

It can be shown that

$$-1 \leq \rho(X, Y) \leq 1.$$

If X and Y are linearly related by the equation

$$Y = a + bX,$$

then $\rho(X, Y)$ will equal 1 when b is positive and -1 when b is negative.

1.5   Exercises

Exercise 1.1   When typing a report, a certain typist makes i errors with probability p_i $(i \geq 0)$, where

$$p_0 = .20, \quad p_1 = .35, \quad p_2 = .25, \quad p_3 = .15.$$

What is the probability that the typist makes

(a) at least four errors;
(b) at most two errors?

Exercise 1.2 A family picnic scheduled for tomorrow will be post-poned if it is either cloudy or rainy. If the probability that it will be cloudy is .40, the probability that it will be rainy is .30, and the proba-bility that it will be both rainy and cloudy is .20, what is the probabilty that the picnic will not be postponed?

Exercise 1.3 If two people are randomly chosen from a group of eight women and six men, what is the probability that

(a) both are women;
(b) both are men;
(c) one is a man and the other a woman?

Exercise 1.4 A club has 120 members, of whom 35 play chess, 58 play bridge, and 27 play both chess and bridge. If a member of the club is randomly chosen, what is the conditional probability that she

(a) plays chess given that she plays bridge;
(b) plays bridge given that she plays chess?

Exercise 1.5 Cystic fibrosis (CF) is a genetically caused disease. A child that receives a CF gene from each of its parents will develop the disease either as a teenager or before, and will not live to adulthood. A child that receives either zero or one CF gene will not develop the dis-ease. If an individual has a CF gene, then each of his or her children will independently receive that gene with probability $1/2$.

(a) If both parents possess the CF gene, what is the probability that their child will develop cystic fibrosis?
(b) What is the probability that a 30-year old who does not have cys-tic fibrosis, but whose sibling died of that disease, possesses a CF gene?

Exercise 1.6 Two cards are randomly selected from a deck of 52 playing cards. What is the conditional probability they are both aces, given that they are of different suits?

Exercise 1.7 If A and B are independent, show that so are

(a) A and B^c;
(b) A^c and B^c.

Exercise 1.8 A gambling book recommends the following strategy for the game of roulette. It recommends that the gambler bet 1 on red. If red appears (which has probability 18/38 of occurring) then the gambler should take his profit of 1 and quit. If the gambler loses this bet, he should then make a second bet of size 2 and then quit. Let X denote the gambler's winnings.

(a) Find $P\{X > 0\}$.
(b) Find $E[X]$.

Exercise 1.9 Four buses carrying 152 students from the same school arrive at a football stadium. The buses carry (respectively) 39, 33, 46, and 34 students. One of the 152 students is randomly chosen. Let X denote the number of students who were on the bus of the selected student. One of the four bus drivers is also randomly chosen. Let Y be the number of students who were on that driver's bus.

(a) Which do you think is larger, $E[X]$ or $E[Y]$?
(b) Find $E[X]$ and $E[Y]$.

Exercise 1.10 Two players play a tennis match, which ends when one of the players has won two sets. Suppose that each set is equally likely to be won by either player, and that the results from different sets are independent. Find (a) the expected value and (b) the variance of the number of sets played.

Exercise 1.11 Verify that

$$\text{Var}(X) = E[X^2] - (E[X])^2.$$

Hint: Starting with the definition

$$\text{Var}(X) = E[(X - E[X])^2],$$

square the expression on the right side; then use the fact that the expected value of a sum of random variables is equal to the sum of their expectations.

Exercise 1.12   A lawyer must decide whether to charge a fixed fee of $5,000 or take a contingency fee of $25,000 if she wins the case (and 0 if she loses). She estimates that her probability of winning is .30. Determine the mean and standard deviation of her fee if

(a) she takes the fixed fee;
(b) she takes the contingency fee.

Exercise 1.13   Let $X_1, \ldots, X_n$ be independent random variables, all having the same distribution with expected value μ and variance σ^2. The random variable $\bar{X}$, defined as the arithmetic average of these variables, is called the *sample mean*. That is, the sample mean is given by

$$\bar{X} = \frac{\sum_{i=1}^{n} X_i}{n}.$$

(a) Show that $E[\bar{X}] = \mu$.
(b) Show that $\text{Var}(\bar{X}) = \sigma^2/n$.

The random variable S^2, defined by

$$S^2 = \frac{\sum_{i=1}^{n}(X_i - \bar{X})^2}{n - 1},$$

is called the *sample variance*.

(c) Show that $\sum_{i=1}^{n}(X_i - \bar{X})^2 = \sum_{i=1}^{n} X_i^2 - n\bar{X}^2$.
(d) Show that $E[S^2] = \sigma^2$.

Exercise 1.14   Verify that

$$\text{Cov}(X, Y) = E[XY] - E[X]E[Y].$$

Exercise 1.15   Prove:

(a) $\mathrm{Cov}(X, Y) = \mathrm{Cov}(Y, X)$;
(b) $\mathrm{Cov}(X, X) = \mathrm{Var}(X)$;
(c) $\mathrm{Cov}(cX, Y) = c\,\mathrm{Cov}(X, Y)$;
(d) $\mathrm{Cov}(c, Y) = 0$.

Exercise 1.16   If U and V are independent random variables, both having variance 1, find $\mathrm{Cov}(X, Y)$ when

$$X = aU + bV, \qquad Y = cU + dV.$$

Exercise 1.17   If $\mathrm{Cov}(X_i, X_j) = ij$, find

(a) $\mathrm{Cov}(X_1 + X_2, X_3 + X_4)$;
(b) $\mathrm{Cov}(X_1 + X_2 + X_3, X_2 + X_3 + X_4)$.

Exercise 1.18   Suppose that – in any given time period – a certain stock is equally likely to go up 1 unit or down 1 unit, and that the outcomes of different periods are independent. Let X be the amount the stock goes up (either 1 or -1) in the first period, and let Y be the cumulative amount it goes up in the first three periods. Find the correlation between X and Y.

Exercise 1.19   Can you construct a pair of random variables such that $\mathrm{Var}(X) = \mathrm{Var}(Y) = 1$ and $\mathrm{Cov}(X, Y) = 2$?

REFERENCE

[1] Ross, S. M. (1997). *A First Course in Probability,* 5th ed. Englewood Cliffs, NJ: Prentice-Hall.

2. Normal Random Variables

2.1 Continuous Random Variables

Whereas the possible values of the random variables considered in the previous chapter constituted sets of discrete values, there exist random variables whose set of possible values is instead a continuous region. These *continuous* random variables can take on any value within some interval. For example, such random variables as the time it takes to complete an assignment, or the weight of a randomly chosen individual, are usually considered to be continuous.

Every continuous random variable X has a function f associated with it. This function, called the *probability density function* of X, determines the probabilities associated with X in the following manner. For any numbers $a < b$, the area under f between a and b is equal to the probability that X assumes a value between a and b. That is,

$$P\{a \leq X \leq b\} = \text{area under } f \text{ between } a \text{ and } b.$$

Figure 2.1 presents a probability density function.

2.2 Normal Random Variables

A very important type of continuous random variable is the normal random variable. The probability density function of a normal random variable X is determined by two parameters, denoted by μ and σ, and is given by the formula

$$f(x) = \frac{1}{\sqrt{2\pi}\sigma} e^{-(x-\mu)^2/2\sigma^2}, \quad -\infty < x < \infty.$$

A plot of the normal probability density function gives a bell-shaped curve that is symmetric about the value μ, and with a variability that is measured by σ. The larger the value of σ, the more spread there is in f. Figure 2.2 presents three different normal probability density functions. Note how the curve flattens out as σ increases.

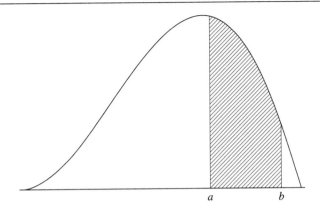

$P\{a \le X \le b\}$ = area of shaded region

Figure 2.1: Probability Density Function of X

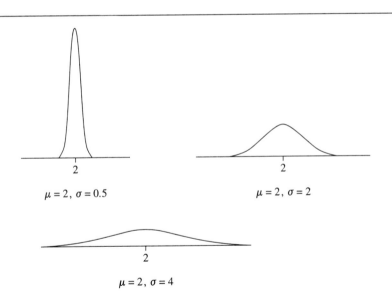

$\mu = 2, \sigma = 0.5$ $\mu = 2, \sigma = 2$

$\mu = 2, \sigma = 4$

Figure 2.2: Three Normal Probability Density Functions

It can be shown that the parameters μ and σ^2 are equal to the expected value and to the variance of X, respectively. That is,

$$\mu = E[X], \qquad \sigma^2 = \text{Var}(X).$$

A normal random variable having mean 0 and variance 1 is called a *standard normal* random variable. Let Z be a standard normal random variable. The function $\Phi(x)$, defined for all real numbers x by

$$\Phi(x) = P\{Z \le x\},$$

is called the *standard normal distribution function*. Thus $\Phi(x)$, the probability that a standard normal random variable is less than or equal to x, is equal to the area under the *standard normal density function*

$$f(x) = \frac{1}{\sqrt{2\pi}} e^{-x^2/2}, \qquad -\infty < x < \infty,$$

between $-\infty$ and x. Table 2.1 specifies values of $\Phi(x)$ when $x > 0$. Probabilities for negative x can be obtained by using the symmetry of the standard normal density about 0 to conclude (see Figure 2.3) that

$$P\{Z < -x\} = P\{Z > x\}$$

or, equivalently, that

$$\Phi(-x) = 1 - \Phi(x).$$

Example 2.2a Let Z be a standard normal random variable. For $a < b$, express $P\{a < Z \le b\}$ in terms of Φ.

Solution. Since

$$P\{Z \le b\} = P\{Z \le a\} + P\{a < Z \le b\},$$

we see that

$$P\{a < Z \le b\} = \Phi(b) - \Phi(a). \qquad \Box$$

Example 2.2b Tabulated values of $\Phi(x)$ show that, to four decimal places,

$$P\{|Z| \le 1\} = P\{-1 \le Z \le 1\} = .6826,$$
$$P\{|Z| \le 2\} = P\{-2 \le Z \le 2\} = .9544,$$
$$P\{|Z| \le 3\} = P\{-3 \le Z \le 3\} = .9974. \qquad \Box$$

Table 2.1: $\Phi(x) = P\{Z \le x\}$

x	.00	.01	.02	.03	.04	.05	.06	.07	.08	.09
0.0	.5000	.5040	.5080	.5120	.5160	.5199	.5239	.5279	.5319	.5359
0.1	.5398	.5438	.5478	.5517	.5557	.5596	.5636	.5675	.5714	.5753
0.2	.5793	.5832	.5871	.5910	.5948	.5987	.6026	.6064	.6103	.6141
0.3	.6179	.6217	.6255	.6293	.6331	.6368	.6406	.6443	.6480	.6517
0.4	.6554	.6591	.6628	.6664	.6700	.6736	.6772	.6808	.6844	.6879
0.5	.6915	.6950	.6985	.7019	.7054	.7088	.7123	.7157	.7190	.7224
0.6	.7257	.7291	.7324	.7357	.7389	.7422	.7454	.7486	.7517	.7549
0.7	.7580	.7611	.7642	.7673	.7704	.7734	.7764	.7794	.7823	.7852
0.8	.7881	.7910	.7939	.7967	.7995	.8023	.8051	.8078	.8106	.8133
0.9	.8159	.8186	.8212	.8238	.8264	.8289	.8315	.8340	.8365	.8389
1.0	.8413	.8438	.8461	.8485	.8508	.8531	.8554	.8577	.8599	.8621
1.1	.8643	.8665	.8686	.8708	.8729	.8749	.8770	.8790	.8810	.8830
1.2	.8849	.8869	.8888	.8907	.8925	.8944	.8962	.8980	.8997	.9015
1.3	.9032	.9049	.9066	.9082	.9099	.9115	.9131	.9147	.9162	.9177
1.4	.9192	.9207	.9222	.9236	.9251	.9265	.9279	.9292	.9306	.9319
1.5	.9332	.9345	.9357	.9370	.9382	.9394	.9406	.9418	.9429	.9441
1.6	.9452	.9463	.9474	.9484	.9495	.9505	.9515	.9525	.9535	.9545
1.7	.9554	.9564	.9573	.9582	.9591	.9599	.9608	.9616	.9625	.9633
1.8	.9641	.9649	.9656	.9664	.9671	.9678	.9686	.9693	.9699	.9706
1.9	.9713	.9719	.9726	.9732	.9738	.9744	.9750	.9756	.9761	.9767
2.0	.9772	.9778	.9783	.9788	.9793	.9798	.9803	.9808	.9812	.9817
2.1	.9821	.9826	.9830	.9834	.9838	.9842	.9846	.9850	.9854	.9857
2.2	.9861	.9864	.9868	.9871	.9875	.9878	.9881	.9884	.9887	.9890
2.3	.9893	.9896	.9898	.9901	.9904	.9906	.9909	.9911	.9913	.9916
2.4	.9918	.9920	.9922	.9925	.9927	.9929	.9931	.9932	.9934	.9936
2.5	.9938	.9940	.9941	.9943	.9945	.9946	.9948	.9949	.9951	.9952
2.6	.9953	.9955	.9956	.9957	.9959	.9960	.9961	.9962	.9963	.9964
2.7	.9965	.9966	.9967	.9968	.9969	.9970	.9971	.9972	.9973	.9974
2.8	.9974	.9975	.9976	.9977	.9977	.9978	.9979	.9979	.9980	.9981
2.9	.9981	.9982	.9982	.9983	.9984	.9984	.9985	.9985	.9986	.9986
3.0	.9987	.9987	.9987	.9988	.9988	.9989	.9989	.9989	.9990	.9990
3.1	.9990	.9991	.9991	.9991	.9992	.9992	.9992	.9992	.9993	.9993
3.2	.9993	.9993	.9994	.9994	.9994	.9994	.9994	.9995	.9995	.9995
3.3	.9995	.9995	.9995	.9996	.9996	.9996	.9996	.9996	.9996	.9997
3.4	.9997	.9997	.9997	.9997	.9997	.9997	.9997	.9997	.9997	.9998

When greater accuracy than that provided by Table 2.1 is needed, the following approximation to $\Phi(x)$, accurate to six decimal places, can be used: For $x > 0$,

$$\Phi(x) \approx 1 - \frac{1}{\sqrt{2\pi}} e^{-x^2/2}(a_1 y + a_2 y^2 + a_3 y^3 + a_4 y^4 + a_5 y^5),$$

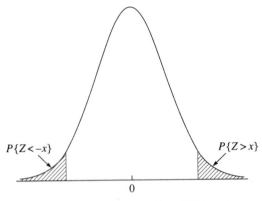

Figure 2.3: $P\{Z < -x\} = P\{Z > x\}$

where

$$y = \frac{1}{1 + .2316419x},$$

$$a_1 = .319381530,$$

$$a_2 = -.356563782,$$

$$a_3 = 1.781477937,$$

$$a_4 = -1.821255978,$$

$$a_5 = 1.330274429,$$

and

$$\Phi(-x) = 1 - \Phi(x).$$

2.3   Properties of Normal Random Variables

An important property of normal random variables is that if X is a normal random variable then so is $aX + b$, when a and b are constants. This property enables us to transform any normal random variable X into a standard normal random variable. For suppose X is normal with mean μ and variance σ^2. Then, since (from Equations (1.7) and (1.8))

$$Z = \frac{X - \mu}{\sigma}$$

has expected value 0 and variance 1, it follows that Z is a standard normal random variable. As a result, we can compute probabilities for any normal random variable in terms of the standard normal distribution function Φ.

Example 2.3a IQ examination scores for sixth-graders are normally distributed with mean value 100 and standard deviation 14.2. What is the probability that a randomly chosen sixth-grader has an IQ score greater than 130?

Solution. Let X be the score of a randomly chosen sixth-grader. Then,

$$P\{X > 130\} = P\left\{\frac{X - 100}{14.2} > \frac{130 - 100}{14.2}\right\}$$

$$= P\left\{\frac{X - 100}{14.2} > 2.113\right\}$$

$$= 1 - \Phi(2.113)$$

$$= .017. \qquad \square$$

Example 2.3b Let X be a normal random variable with mean μ and standard deviation σ. Then, since

$$X - \mu \leq a\sigma$$

is equivalent to

$$\frac{X - \mu}{\sigma} \leq a,$$

it follows from Example 2.2b that 68.26% of the time a normal random variable will be within one standard deviation of its mean; 95.44% of the time it will be within two standard deviations of its mean; and 99.74% of the time it will be within three standard deviations of its mean. $\square$

Another important property of normal random variables is that the sum of independent normal random variables is also a normal random variable. That is, if X_1 and X_2 are independent normal random variables with means μ_1 and μ_2 and with standard deviations σ_1 and σ_2, then $X_1 + X_2$ is normal with mean

$$E[X_1 + X_2] = E[X_1] + E[X_2] = \mu_1 + \mu_2$$

and variance

$$\text{Var}(X_1 + X_2) = \text{Var}(X_1) + \text{Var}(X_2) = \sigma_1^2 + \sigma_2^2.$$

Example 2.3c The annual rainfall in Cleveland, Ohio, is normally distributed with mean 40.14 inches and standard deviation 8.7 inches. Find the probabiity that the sum of the next two years' rainfall exceeds 84 inches.

Solution. Let X_i denote the rainfall in year i ($i = 1, 2$). Then, assuming that the rainfalls in successive years can be assumed to be independent, it follows that $X_1 + X_2$ is normal with mean 80.28 and variance $2(8.7)^2 = 151.38$. Therefore, with Z denoting a standard normal random variable,

$$P\{X_1 + X_2 > 84\} = P\left\{ Z > \frac{84 - 80.28}{\sqrt{151.38}} \right\}$$

$$= P\{Z > .3023\}$$

$$\approx .3812. \qquad \square$$

The random variable Y is said to be a *lognormal* random variable with parameters μ and σ if $\log(Y)$ is a normal random variable with mean μ and variance σ^2. That is, Y is lognormal if it can be expressed as

$$Y = e^X,$$

where X is a normal random variable. The mean and variance of a lognormal random variable are as follows:

$$E[Y] = e^{\mu + \sigma^2/2},$$

$$\text{Var}(Y) = e^{2\mu + 2\sigma^2} - e^{2\mu + \sigma^2} = e^{2\mu + \sigma^2}(e^{\sigma^2} - 1).$$

Example 2.3d Starting at some fixed time, let $S(n)$ denote the price of a certain security at the end of n additional weeks, $n \geq 1$. A popular model for the evolution of these prices assumes that the price ratios $S(n)/S(n-1)$ for $n \geq 1$ are independent and identically distributed (i.i.d.) lognormal random variables. Assuming this model, with lognormal parameters $\mu = .0165$ and $\sigma = .0730$, what is the probability that

(a) the price of the security increases over each of the next two weeks;
(b) the price at the end of two weeks is higher than it is today?

Solution. Let Z be a standard normal random variable. To solve part (a), we use that $\log(x)$ increases in x to conclude that $x > 1$ if and only if $\log(x) > \log(1) = 0$. As a result, we have

$$P\left\{\frac{S(1)}{S(0)} > 1\right\} = P\left\{\log\left(\frac{S(1)}{S(0)}\right) > 0\right\}$$

$$= P\left\{Z > \frac{-.0165}{.0730}\right\}$$

$$= P\{Z > -.2260\}$$

$$= P\{Z < .2260\}$$

$$\approx .5894.$$

Therefore, the probability that the price is up after one week is .5894. Since the successive price ratios are independent, the probability that the price increases over each of the next two weeks is $(.5894)^2 = .3474$.

To solve part (b), reason as follows:

$$P\left\{\frac{S(2)}{S(0)} > 1\right\} = P\left\{\frac{S(2)}{S(1)}\frac{S(1)}{S(0)} > 1\right\}$$

$$= P\left\{\log\left(\frac{S(2)}{S(1)}\right) + \log\left(\frac{S(1)}{S(0)}\right) > 0\right\}$$

$$= P\left\{Z > \frac{-.0330}{.0730\sqrt{2}}\right\}$$

$$= P\{Z > -.31965\}$$

$$= P\{Z < .31965\}$$

$$\approx .6254,$$

where we have used that $\log\left(\frac{S(2)}{S(1)}\right) + \log\left(\frac{S(1)}{S(0)}\right)$, being the sum of independent normal random variables with a common mean .0165 and a common standard deviation .0730, is itself a normal random variable with mean .0330 and variance $2(.0730)^2$. $\qquad\square$

2.4 The Central Limit Theorem

The ubiquity of normal random variables is explained by the central limit theorem, probably the most important theoretical result in probability.

This theorem states that the sum of a large number of independent random variables, all having the same probability distribution, will itself be approximately a normal random variable.

For a more precise statement of the central limit theorem, suppose that $X_1, X_2, \ldots$ is a sequence of i.i.d. random variables, each with expected value μ and variance σ^2, and let

$$S_n = \sum_{i=1}^{n} X_i.$$

Central Limit Theorem   *For large n, S_n will approximately be a normal random variable with expected value $n\mu$ and variance $n\sigma^2$. As a result, for any x we have*

$$P\left\{ \frac{S_n - n\mu}{\sigma\sqrt{n}} \leq x \right\} \approx \Phi(x),$$

with the approximation becoming exact as n becomes larger and larger.

Suppose that X is a binomial random variable with parameters n and p. Since X represents the number of successes in n independent trials, each of which is a success with probability p, it can be expressed as

$$X = \sum_{i=1}^{n} X_i,$$

where X_i is 1 if trial i is a success and is 0 otherwise. Since (from Section 1.3)

$$E[X_i] = p \quad \text{and} \quad \text{Var}(X_i) = p(1-p),$$

it follows from the central limit theorem that, when n is large, X will approximately have a normal distribution with mean np and variance $np(1-p)$.

Example 2.4a   A fair coin is tossed 100 times. What is the probability that heads appears fewer than 40 times?

Solution. If X denotes the number of heads, then X is a binomial random variable with parameters $n = 100$ and $p = 1/2$. Since $np = 50$ we have $np(1-p) = 25$, and so

$$P\{X < 40\} = P\left\{ \frac{X - 50}{\sqrt{25}} < \frac{40 - 50}{\sqrt{25}} \right\}$$

$$= P\left\{ \frac{X - 50}{\sqrt{25}} < -2 \right\}$$

$$\approx \Phi(-2)$$

$$= .0228.$$

A computer program for computing binomial probabilities gives the exact solution .0176, and so the preceding is not quite as acccurate as we might like. However, we could improve the approximation by noting that, since X is an integral-valued random variable, the event that $X < 40$ is equivalent to the event that $X < 39 + c$ for any c, $0 < c \leq 1$. Consequently, a better approximation may be obtained by writing the desired probability as $P\{X < 39.5\}$. This gives

$$P\{X < 39.5\} = P\left\{ \frac{X - 50}{\sqrt{25}} < \frac{39.5 - 50}{\sqrt{25}} \right\}$$

$$= P\left\{ \frac{X - 50}{\sqrt{25}} < -2.1 \right\}$$

$$\approx \Phi(-2.1)$$

$$= .0179,$$

which is indeed a better approximation. □

2.5   Exercises

Exercise 2.1   For a standard normal random variable Z, find:

(a) $P\{Z < -.66\}$;
(b) $P\{|Z| < 1.64\}$;
(c) $P\{|Z| > 2.20\}$.

Exercise 2.2   Find the value of x when Z is a standard normal random variable and

$$P\{-2 < Z < -1\} = P\{1 < Z < x\}.$$

Exercise 2.3   Argue (a picture is acceptable) that

$$P\{|Z| > x\} = 2P\{Z > x\},$$

where $x > 0$ and Z is a standard normal random variable.

Exercise 2.4   Let X be a normal random variable having expected value μ and variance σ^2, and let $Y = a + bX$. Find values a, b ($a \neq 0$) that give Y the same distribution as X. Then, using these values, find $\text{Cov}(X, Y)$.

Exercise 2.5   The systolic blood pressure of male adults is normally distributed with a mean of 127.7 and a standard deviation of 19.2.

(a) Specify an interval in which the blood pressures of approximately 68% of the adult male population fall.
(b) Specify an interval in which the blood pressures of approximately 95% of the adult male population fall.
(c) Specify an interval in which the blood pressures of approximately 99.7% of the adult male population fall.

Exercise 2.6   Suppose that the amount of time that a certain battery functions is a normal random variable with mean 400 hours and standard deviation 50 hours. Suppose that an individual owns two such batteries, one of which is to be used as a spare to replace the other when it fails.

(a) What is the probability that the total life of the batteries will exceed 760 hours?
(b) What is the probability that the second battery will outlive the first by at least 25 hours?
(c) What is the probability that the longer-lasting battery will outlive the other by at least 25 hours?

Exercise 2.7   The time it takes to develop a photographic print is a random variable with mean 18 seconds and standard deviation 1 second. Approximate the probability that the total amount of time that it takes to process 100 prints is

(a) more than 1,710 seconds;
(b) between 1,690 and 1,710 seconds.

Exercise 2.8   Frequent fliers of a certain airline fly a random number of miles each year, having mean and standard deviation of 25,000 and 12,000 miles, respectively. If 30 such people are randomly chosen, approximate the probability that the average of their mileages for this year will

(a) exceed 25,000;
(b) be between 23,000 and 27,000.

Exercise 2.9   A model for the movement of a stock supposes that, if the present price of the stock is s, then – after one time period – it will either be us with probability p or ds with probability $1 - p$. Assuming that successive movements are independent, approximate the probability that the stock's price will be up at least 30% after the next 1,000 time periods if $u = 1.012$, $d = .990$, and $p = .52$.

Exercise 2.10   In each time period, a certain stock either goes down 1 with probability .39, remains the same with probability .20, or goes up 1 with probability .41. Asuming that the changes in successive time periods are independent, approximate the probability that, after 700 time periods, the stock will be up more than 10 from where it started.

3. Geometric Brownian Motion

3.1 Geometric Brownian Motion

Suppose that we are interested in the price of some security as it evolves over time. Let the present time be time 0, and let $S(y)$ denote the price of the security a time y from the present. We say that the collection of prices $S(y)$, $0 \leq y < \infty$, follows a *geometric Brownian motion* with drift parameter μ and volatility parameter σ if, for all nonegative values of y and t, the random variable

$$\frac{S(t+y)}{S(y)}$$

is independent of all prices up to time y; and if, in addition,

$$\log\left(\frac{S(t+y)}{S(y)}\right)$$

is a normal random variable with mean μt and variance $t\sigma^2$.

In other words, the series of prices will be a geometric Brownian motion if the ratio of the price a time t in the future to the present price will, independent of the past history of prices, have a lognormal probability distribution with parameters μt and $t\sigma^2$.

It follows that a consequence of assuming a security's prices follow a geometric Brownian motion is that, once μ and σ are determined, it is only the present price – and not the history of past prices – that affects probabilities of future prices. Furthermore, probabilities concerning the ratio of the price a time t in the future to the present price will not depend on the present price. (Thus, for instance, the model implies that the probability a given security doubles in price in the next month is the same no matter whether its present price is 10 or 25.)

It turns out that, for a given initial price $S(0)$, the expected value of the price at time t depends on *both* of the geometric Brownian motion parameters. Specifically, if the initial price is s_0, then

$$E[S(t)] = s_0 e^{t(\mu + \sigma^2/2)}.$$

Thus, under geometric Brownian motion, the expected price grows at the rate $\mu + \sigma^2/2$.

3.2 Geometric Brownian Motion as a Limit of Simpler Models

Let Δ denote a small increment of time and suppose that, every Δ time units, the price of a security either goes up by the factor u with probability p or goes down by the factor d with probability $1 - p$, where

$$u = e^{\sigma\sqrt{\Delta}}, \qquad d = e^{-\sigma\sqrt{\Delta}},$$

$$p = \frac{1}{2}\left(1 + \frac{\mu}{\sigma}\sqrt{\Delta}\right).$$

That is, we are supposing that the price of the security changes only at times that are integral multiples of Δ; at these times, it either goes up by the factor u or down by the factor d.

As we take Δ smaller and smaller, so that the price changes occur more and more frequently (though by factors that become closer and closer to 1), the collection of prices becomes a geometric Brownian motion. Consequently, geometric Brownian motion can be approximated by a relatively simple process, one that goes either up or down by fixed factors at regularly specified times.

Let us now verify that the preceding model becomes geometric Brownian motion as we let Δ become smaller and smaller. To begin, let Y_i equal 1 if the price goes up at time $i\Delta$, and let it be 0 if it goes down. Now, the number of times that the security's price goes up in the first n time increments is $\sum_{i=1}^{n} Y_i$, and the number of times it goes down is $n - \sum_{i=1}^{n} Y_i$. Hence, $S(n\Delta)$, its price at the end of this time, can be expressed as

$$S(n\Delta) = S(0)u^{\sum_{i=1}^{n} Y_i} d^{n - \sum_{i=1}^{n} Y_i}$$

or

$$S(n\Delta) = d^n S(0)\left(\frac{u}{d}\right)^{\sum_{i=1}^{n} Y_i}.$$

If we now let $n = t/\Delta$, then the preceding equation can be expressed as

$$\frac{S(t)}{S(0)} = d^{t/\Delta} \left(\frac{u}{d}\right)^{\sum_{i=1}^{t/\Delta} Y_i}.$$

Taking logarithms gives

$$\log\left(\frac{S(t)}{S(0)}\right) = \frac{t}{\Delta} \log(d) + \log\left(\frac{u}{d}\right) \sum_{i=1}^{t/\Delta} Y_i$$

$$= \frac{-t\sigma}{\sqrt{\Delta}} + 2\sigma\sqrt{\Delta} \sum_{i=1}^{t/\Delta} Y_i, \qquad (3.1)$$

where Equation (3.1) used the definitions of u and d. Now, as Δ goes to 0, there are more and more terms in the summation $\sum_{i=1}^{t/\Delta} Y_i$; hence, by the central limit theorem, this sum becomes more and more normal, implying from Equation (3.1) that $\log(S(t)/S(0))$ becomes a normal random variable. Moreover, from Equation (3.1) we obtain that

$$E\left[\log\left(\frac{S(t)}{S(0)}\right)\right] = \frac{-t\sigma}{\sqrt{\Delta}} + 2\sigma\sqrt{\Delta} \sum_{i=1}^{t/\Delta} E[Y_i]$$

$$= \frac{-t\sigma}{\sqrt{\Delta}} + 2\sigma\sqrt{\Delta}\frac{t}{\Delta}p$$

$$= \frac{-t\sigma}{\sqrt{\Delta}} + \frac{t\sigma}{\sqrt{\Delta}}\left(1 + \frac{\mu}{\sigma}\sqrt{\Delta}\right)$$

$$= \mu t.$$

Furthermore, Equation (3.1) yields that

$$\text{Var}\left(\log\left(\frac{S(t)}{S(0)}\right)\right) = 4\sigma^2\Delta \sum_{i=1}^{t/\Delta} \text{Var}(Y_i) \qquad \text{(by independence)}$$

$$= 4\sigma^2 tp(1-p)$$

$$\approx \sigma^2 t \qquad \text{(since, for small } \Delta, \ p \approx 1/2\text{)}.$$

Thus we see that, as Δt becomes smaller and smaller, $\log(S(t)/S(0))$ (and, by the same reasoning, $\log(S(t+y)/S(y)))$ becomes a normal random variable with mean μt and variance $t\sigma^2$. In addition, because successive price changes are independent and each has the same probability of being an increase, it follows that $S(t+y)/S(y)$ is independent of earlier price changes before time y. Hence, as Δ goes to 0, both conditions of geometric Brownian motion are met, showing that the model indeed becomes geometric Brownian motion.

3.3 Brownian Motion

Geometric Brownian motion can be considered to be a variant of a long-studied model known as Brownian motion. It is defined as follows.

Definition The collection of prices $S(y)$, $0 \leq y < \infty$, is said to follow a *Brownian motion* with drift parameter μ and variance parameter σ^2 if, for all nonegative values of y and t, the random variable

$$S(t + y) - S(y)$$

is independent of all prices up to time y and, in addition, is a normal random variable with mean μt and variance $t\sigma^2$.

Thus, Brownian motion shares with geometric Brownian motion the property that a future price depends on the present and all past prices only through the present price; however, in Brownian motion it is the difference in prices (and not the logarithm of their ratio) that has a normal distribution.

The Brownian motion process has an distinguished scientific pedigree. It is named after the English botanist Robert Brown, who first described (in 1827) the unusual motion exhibited by a small particle that is totally immersed in a liquid or gas. The first explanation of this motion was given by Albert Einstein in 1905. He showed mathematically that Brownian motion could be explained by assuming that the immersed particle was continually being subjected to bombardment by the molecules of the surrounding medium. A mathematically concise definition, as well as an elucidation of some of the mathematical properties of Brownian motion, was given by the American applied mathematician Norbert Wiener in a series of papers originating in 1918.

Interestingly, Brownian motion was independently introduced in 1900 by the French mathematician Bachelier, who used it in his doctoral dissertation to model the price movements of stocks and commodities. However, Brownian motion appears to have two major flaws when used to model stock or commodity prices. First, since the price of a stock is a normal random variable, it can theoretically become negative. Second, the assumption that a price *difference* over an interval of fixed length has the same normal distribution no matter what the price at the beginning of the interval does not seem totally reasonable. For instance, many people might not think that the probability a stock presently selling at $20 would drop to $15 (a loss of 25%) in one month would be the same as the probability that when the stock is at $10 it would drop to $5 (a loss of 50%) in one month.

The geometric Brownian motion model, on the other hand, possesses neither of these flaws. Since it is now the *logarithm* of the stock's price that is a normal random variable, the model does not allow for negative stock prices. In addition, since it is the ratios of prices separated by a fixed length of time that have the same distribution, geometric Brownian motion makes what many feel is the more reasonable assumption that it is the *percentage* change in price, and not the absolute change, whose probabilities do not depend on the present price. However, it should be noted that – in both of these models – once the model parameters μ and σ are determined, the only information that is needed for predicting future prices is the present price; information about past prices is irrelevant.

3.4   Exercises

Exercise 3.1   Suppose that $S(y)$, $y \geq 0$, is a geometric Brownian motion with drift parameter $\mu = .01$ and volatility parameter $\sigma = .2$. If $S(0) = 100$, find:

(a) $E[S(10)]$;
(b) $P\{S(10) > 100\}$;
(c) $P\{S(10) < 110\}$.

Exercise 3.2   Repeat Exercise 3.1 when the volatility parameter is equal to .4.

Exercise 3.3   Repeat Exercise 3.2 when the volatility parameter is equal to .6.

Exercise 3.4 It can be shown that if X is a normal random variable with mean m and variance v^2, then

$$E[e^X] = e^{m+v^2/2}.$$

Use this result to verify the formula for $E[S(t)]$ given in Section 3.1.

Exercise 3.5 Use the result of the preceding exercise to find $\text{Var}(S(t))$ when $S(0) = s_0$.

Hint: Use the identity

$$\text{Var}(X) = E[X^2] - (E[X])^2.$$

REFERENCES

[1] Bachelier, Louis (1900). "Theorie de la Speculation." *Annales de l'École Normale Supérieure* 17: 21–86; English translation by A. J. Boness in P. H. Cootner (Ed.) (1964), *The Random Character of Stock Market Prices,* pp. 17–78. Cambridge, MA: MIT Press.
[2] Ross, S. M. (1997). *Introduction To Probability Models,* 6th ed. Orlando, FL: Academic Press.

4. Interest Rates and Present Value Analysis

4.1 Interest Rates

If you borrow the amount P (called the principal), which must be repaid after a time T along with simple interest at rate r per time T, then the amount to be repaid at time T is

$$P + rP = P(1+r).$$

That is, you must repay both the principal P and the interest, equal to the principal times the interest rate. For instance, if you borrow $100 to be repaid after one year with a simple interest rate of 5% per year (i.e., $r = .05$), then you will have to repay $105 at the end of the year.

Example 4.1a Suppose that you borrow the amount P, to be repaid after one year along with interest at a rate r per year *compounded* semiannually. What does this mean? How much is owed in a year?

Solution. In order to solve this example, you must realize that having your interest compounded semiannually means that after half a year you are to be charged simple interest at the rate of $r/2$ per half-year, and that interest is then added on to your principal, which is again charged interest at rate $r/2$ for the second half-year period. In other words, after six months you owe

$$P(1 + r/2).$$

This is then regarded as the new principal for another six-month loan at interest rate $r/2$; hence, at the end of the year you will owe

$$P(1 + r/2)(1 + r/2) = P(1 + r/2)^2. \qquad \Box$$

Example 4.1b If you borrow $1,000 for one year at an interest rate of 8% per year compounded quarterly, how much do you owe at the end of the year?

Solution. An interest rate of 8% that is compounded quarterly is equivalent to paying simple interest at 2% per quarter-year, with each successive quarter charging interest not only on the original principal but also on the interest that has accrued up to that point. Thus, after one quarter you owe

$$1,000(1 + .02);$$

after two quarters you owe

$$1,000(1 + .02)(1 + .02) = 1,000(1 + .02)^2;$$

after three quarters you owe

$$1,000(1 + .02)^2(1 + .02) = 1,000(1 + .02)^3;$$

and after four quarters you owe

$$1,000(1 + .02)^3(1 + .02) = 1,000(1 + .02)^4 = \$1,082.40. \qquad \square$$

Example 4.1c   Many credit-card companies charge interest at a yearly rate of 18% compounded monthly. If the amount P is charged at the beginning of a year, how much is owed at the end of the year if no previous payments have been made?

Solution. Such a compounding is equivalent to paying simple interest every month at a rate of $18/12 = 1.5\%$ per month, with the accrued interest then added to the principal owed during the next month. Hence, after one year you will owe

$$P(1 + .015)^{12} = 1.1956P. \qquad \square$$

If the interest rate r is compounded then, as we have seen in Examples 4.1b and 4.1c, the amount of interest actually paid is greater than if we were paying simple interest at rate r. The reason, of course, is that in compounding we are being charged interest on the interest that has already been computed in previous compoundings. In these cases, we call r the *nominal* interest rate, and we define the *effective interest rate*, call it r_{eff}, by

$$r_{\text{eff}} = \frac{\text{amount repaid at the end of a year} - P}{P}.$$

For instance, if the loan is for one year at a nominal interest rate r that is to be compounded quarterly, then the effective interest rate for the year is

$$r_{eff} = (1 + r/4)^4 - 1.$$

Thus, in Example 4.1b the effective interest rate is 8.24% whereas in Example 4.1c it is 19.56%. Since

$$P(1 + r_{eff}) = \text{amount repaid at the end of a year,}$$

the payment made in a one-year loan with compound interest is the same as if the loan called for simple interest at rate r_{eff} per year.

Suppose now that we borrow the principal P for one year at a nominal interest rate of r per year, compounded *continuously*. Now, how much is owed at the end of the year? Of course, to answer this we must first decide on an appropriate definition of "continuous" compounding. To do so, note that if the loan is compounded at n equal intervals in the year, then the amount owed at the end of the year is $P(1 + r/n)^n$. As it is reasonable to suppose that continuous compounding refers to the limit of this process as n grows larger and larger, the amount owed at time 1 is

$$P \lim_{n \to \infty} (1 + r/n)^n = Pe^r.$$

Example 4.1d   If a bank offers interest at a nominal rate of 5% compounded continuously, what is the effective interest rate per year?

Solution. The effective interest rate is

$$r_{eff} = \frac{Pe^{.05} - P}{P} = e^{.05} - 1 \approx .05127.$$

That is, the effective interest rate is 5.127% per year.   □

If the amount P is borrowed for t years at a nominal interest rate of r per year compounded continuously, then the amount owed at time t is Pe^{rt}. This is seen by interpreting the interest rate as being a continuous compounding of a nominal rate of rt per time t; hence, the amount owed at time t is

$$P \lim_{n \to \infty} (1 + rt/n)^n = Pe^{rt}.$$

Example 4.1e *The Doubling Rule*   If you put funds into an account that pays interest at rate r compounded annually, how many years does it take for your funds to double?

Solution. Since your initial deposit of D will be worth $D(1+r)^n$ after n years, we need to find the value of n such that

$$(1+r)^n = 2.$$

Now,

$$(1+r)^n = \left(1 + \frac{nr}{n}\right)^n$$

$$\approx e^{nr},$$

where the approximation is fairly precise provided that n is not too small. Therefore,

$$e^{nr} \approx 2,$$

implying that

$$n \approx \frac{\log(2)}{r} = \frac{.693}{r}.$$

Thus, it will take n years for your funds to double when

$$n \approx \frac{.7}{r}.$$

For instance, if the interest rate is 1% ($r = .01$) then it will take approximately 70 years for your funds to double; if $r = .02$, it will take about 35 years; if $r = .03$, it will take about $23\frac{1}{3}$ years; if $r = .05$, it will take about 14 years; if $r = .07$, it will take about 10 years; and if $r = .10$, it will take about 7 years.

As a check on the preceding approximations, note that (to three–decimal-place accuracy):

$$(1.01)^{70} = 2.007,$$

$$(1.02)^{35} = 2.000,$$

$$(1.03)^{23.33} = 1.993,$$

$$(1.05)^{14} = 1.980,$$

$$(1.07)^{10} = 1.967,$$

$$(1.10)^{7} = 1.949. \qquad \square$$

4.2   Present Value Analysis

Suppose that one can both borrow and loan money at a nominal rate r that is compounded periodically. Under these conditions, what is the present worth of a payment of v dollars that will be made at the end of period i? Since a bank loan of $v(1+r)^{-i}$ would require a payoff of v at period i, it follows that the *present value* of a payoff of v to be made at time period i is $v(1+r)^{-i}$.

The concept of present value enables us to compare different income streams to see which is preferable.

Example 4.2a   Suppose that you are to receive payments (in thousands of dollars) at the end of each of the next five years. Which of the following three payment sequences is preferable?

A.   12, 14, 16, 18, 20;
B.   16, 16, 15, 15, 15;
C.   20, 16, 14, 12, 10.

Solution. If the nominal interest rate is r compounded yearly, then the present value of the sequence of payments x_i $(i = 1, 2, 3, 4, 5)$ is

$$\sum_{i=1}^{5}(1+r)^{-i}x_i;$$

the sequence having the largest present value is preferred. It thus follows that the superior sequence of payments depends on the interest rate. If r is small, then the sequence **A** is best since its sum of payments is the highest. For a somewhat larger value of r, the sequence **B** would be best because – although the total of its payments (77) is less than that of **A** (80) – its earlier payments are larger than are those of **A**. For an even larger value of r, the sequence **C**, whose earlier payments are higher than those of either **A** or **B**, would be best. Table 4.1 gives the present values of these payment streams for three different values of r.

It should be noted that the payment sequences can be compared according to their values at any specified time. For instance, to compare them in terms of their time-5 values, we would determine which sequence of payments yields the largest value of

Table 4.1: *Present Values*

		Payment Sequence	
r	**A**	**B**	**C**
.1	59.21	58.60	56.33
.2	45.70	46.39	45.69
.3	36.49	37.89	38.12

$$\sum_{i=1}^{5}(1+r)^{5-i}x_i = (1+r)^5\sum_{i=1}^{5}(1+r)^{-i}x_i.$$

Consequently, we obtain the same preference ordering as a function of interest rate as before.   □

Remark. Let $\mathbf{a} = (a_0, a_1, \ldots, a_n)$ and $\mathbf{b} = (b_0, b_1, \ldots, b_n)$ be cash flow sequences, and suppose that the present value of the $\mathbf{a}$ sequence is at least as large as that of the $\mathbf{b}$ sequence when the interest rate is r. That is,

$$PV(\mathbf{a}) = \sum_{i=0}^{n} a_i(1+r)^{-i} \geq \sum_{i=0}^{n} b_i(1+r)^{-i} = PV(\mathbf{b}).$$

One way of seeing the superiority of the $\mathbf{a}$ sequence is to note that it can be transformed, by borrowing and saving at the rate r, into a cash flow sequence $\mathbf{c} = (c_0, c_1, \ldots, c_n)$ having $c_i \geq b_i$ for each $i = 0, \ldots, n$. We prove this fact by induction on n. As it is immediate when $n = 0$, assume that the result holds whenever the cash flow sequences are of length n, and now consider cash flow sequences $\mathbf{a}$ and $\mathbf{b}$ that are of length $n + 1$ and are such that the present value of $\mathbf{a}$ is greater than or equal to that of $\mathbf{b}$. There are two cases to consider.

Case 1: $a_0 \geq b_0$. In this case, start by putting aside the amount b_0 and depositing $a_0 - b_0$ in a bank to be withdrawn in the next period. In this manner, $\mathbf{a}$ is transformed into the cash flow sequence

$$(b_0, (1+r)(a_0 - b_0) + a_1, \ldots, a_n).$$

Now, since

$$(1+r)(a_0 - b_0) + \sum_{i=0}^{n-1} a_{i+1}(1+r)^{-i} - \sum_{i=0}^{n-1} b_{i+1}(1+r)^{-i}$$
$$= (1+r)[PV(\mathbf{a}) - PV(\mathbf{b})] \geq 0,$$

it follows that the time-1 value of the cash flows $(1 + r)(a_0 - b_0) + a_1, \ldots, a_n$ to be received in periods $1, \ldots, n$ is at least as large as that of the cash flows $b_1, \ldots, b_n$. Hence, by the induction hypothesis we can transform the cash flow sequence $(b_0, (1+r)(a_0 - b_0) + a_1, \ldots, a_n)$ into a sequence $(b_0, c_1, \ldots, c_n)$ which is such that $c_i \geq b_i$ for each i. This completes the induction proof in this case.

Case 2: $a_0 < b_0$. In this case, start by borrowing $b_0 - a_0$, to be repaid in period 1. This transforms the cash flow sequence $\mathbf{a}$ into the sequence $(b_0, a_1 - (1+r)(b_0 - a_0), a_2, \ldots, a_n)$. It easily follows that the time-1 value of the cash flows $a_1 - (1+r)(b_0 - a_0), a_2, \ldots, a_n$ to be received at the ends of periods $1, \ldots, n$ is at least as great as that of the cash flows $b_1, \ldots, b_n$, so the result again follows from the induction hypothesis.

Example 4.2b   A company needs a certain type of machine for the next five years. They presently own such a machine, which is now worth $6,000 but will lose $2,000 in value in each of the next three years, after which it will be worthless and unuseable. The (beginning-of-the-year) value of its yearly operating cost is $9,000, with this amount expected to increase by $2,000 in each subsequent year that it is used. A new machine can be purchased at the beginning of any year for a fixed cost of $22,000. The lifetime of a new machine is six years, and its value decreases by $3,000 in each of its first two years of use and then by $4,000 in each following year. The operating cost of a new machine is $6,000 in its first year, with an increase of $1,000 in each subsequent year. If the interest rate is 10%, when should the company purchase a new machine?

Solution. The company can purchase a new machine at the beginning of year 1, 2, 3, or 4, with the following six-year cash flows (in units of $1,000) as a result:

- buy at beginning of year $1 - 22, 7, 8, 9, 10, -4$;
- buy at beginning of year $2 - 9, 24, 7, 8, 9, -8$;
- buy at beginning of year $3 - 9, 11, 26, 7, 8, -12$;
- buy at beginning of year $4 - 9, 11, 13, 28, 7, -16$.

To see why this listing is correct, suppose that the company will buy a new machine at the beginning of year 3. Then its year-1 cost is the $9,000 operating cost of the old machine; its year-2 cost is the $11,000 operating cost of this machine; its year-3 cost is the $22,000 cost of a new machine, plus the $6,000 operating cost of this machine, minus the $2,000 obtained for the replaced machine; its year-4 cost is the $7,000 operating cost; its year-5 cost is the $8,000 operating cost; and its year-6 cost is −$12, 000, the negative of the value of the 3-year-old machine that it no longer needs. The other cash flow sequences are similarly argued.

With the yearly interest rate $r = .10$, the present value of the first cost-flow sequence is

$$22 + \frac{7}{1.1} + \frac{8}{(1.1)^2} + \frac{9}{(1.1)^3} + \frac{10}{(1.1)^4} - \frac{4}{(1.1)^5} = 46.083.$$

The present values of the other cash flows are similarly determined, and the four present values are

$$46.083, \ 43.794, \ 43.760, \ 45.627.$$

Therefore, the company should purchase a new machine two years from now. □

Example 4.2c An individual who plans to retire in 20 years has decided to put an amount A in the bank at the beginning of each of the next 240 months, after which she will withdraw $1,000 at the beginning of each of the following 360 months. Assuming a nominal yearly interest rate of of 6% compounded monthly, how large does A need to be?

Solution. Let $r = .06/12 = .005$ be the monthly interest rate. With $\beta = \frac{1}{1+r}$, the present value of all her deposits is

$$A + A\beta + A\beta^2 + \cdots + A\beta^{239} = A\frac{1 - \beta^{240}}{1 - \beta}.$$

Similarly, if W is the amount withdrawn in the following 360 months, then the present value of all these withdrawals is

$$W\beta^{240} + W\beta^{241} + \cdots + W\beta^{599} = W\beta^{240}\frac{1 - \beta^{360}}{1 - \beta}.$$

Thus she will be able to fund all withdrawals (and have no money left in her account) if

$$A\frac{1-\beta^{240}}{1-\beta} = W\beta^{240}\frac{1-\beta^{360}}{1-\beta}.$$

With $W = 1,000$, and $\beta = 1/1.005$, this gives

$$A = 360.99.$$

That is, saving $361 a month for 240 months will enable her to withdraw $1,000 a month for the succeeding 360 months.

Remark. In this example we have made use of the algebraic identity

$$1 + b + b^2 + \cdots + b^n = \frac{1 - b^{n+1}}{1 - b}.$$

We can prove this identity by letting

$$x = 1 + b + b^2 + \cdots + b^n$$

and then noting that

$$x - 1 = b + b^2 + \cdots + b^n$$
$$= b(1 + b + \cdots + b^{n-1})$$
$$= b(x - b^n).$$

Therefore,

$$(1 - b)x = 1 - b^{n+1},$$

which yields the identity. □

Example 4.2d Suppose you have just spoken to a bank about borrowing $100,000 to purchase a house, and the loan officer has told you that a $100,000 loan, to be repaid in monthly installments over 15 years with an interest rate of .6% per month, could be arranged. If the bank charges a loan initiation fee of $600, a house inspection fee of $400, and 1 "point," what is the effective annual interest rate of the loan being offered?

Solution. To begin, let us determine the monthly mortgage payment, call it A, of such a loan. Since $100,000 is to be repaid in 180 monthly payments at an interest rate of .6% per month, it follows that

$$A[\alpha + \alpha^2 + \cdots + \alpha^{180}] = 100,000,$$

where $\alpha = 1/1.006$. Therefore,

$$A = \frac{100,000(1-\alpha)}{\alpha(1-\alpha^{180})} = 910.05.$$

So if you were actually receiving $100,000 to be repaid in 180 monthly payments of $910.05, then the effective monthly interest rate would be .6%. However, taking into account the initiation and inspection fees involved and the bank charge of 1 point (which means that 1% of the nominal loan of $100,000 must be paid to the bank when the loan is received), it follows that you are actually receiving only $98,000. Consequently, the effective monthly interest rate is that value of r such that

$$A[\beta + \beta^2 + \cdots + \beta^{180}] = 98,000,$$

where $\beta = (1+r)^{-1}$. Therefore,

$$\frac{\beta(1-\beta^{180})}{1-\beta} = 107.69$$

or, since $\frac{1-\beta}{\beta} = r$,

$$\frac{1-\left(\frac{1}{1+r}\right)^{180}}{r} = 107.69.$$

Numerically solving this by trial and error (easily accomplished since we know that $r > .006$) yields the solution

$$r = .00627.$$

Since $(1 + .00627)^{12} = 1.0779$, it follows that what was quoted as a monthly interest rate of .6% is, in reality, an effective annual interest rate of approximately 7.8%. □

Example 4.2e Suppose that one takes a mortgage loan for the amount L that is to be paid back over n months with equal payments of A at the end of each month. The interest rate for the loan is r per month, compounded monthly.

(a) In terms of L, n, and r, what is the value of A?
(b) After payment has been made at the end of month j, how much additional loan principal remains?
(c) How much of the payment during month j is for interest and how much is for principal reduction? (This is important because some contracts allow for the loan to be paid back early and because the interest part of the payment is tax-deductible.)

Solution. The present value of the n monthly payments is

$$\frac{A}{1+r} + \frac{A}{(1+r)^2} + \cdots + \frac{A}{(1+r)^n} = \frac{A}{1+r}\frac{1-\left(\frac{1}{1+r}\right)^n}{1-\frac{1}{1+r}}$$

$$= \frac{A}{r}[1-(1+r)^{-n}].$$

Since this must equal the loan amount L, we see that

$$A = \frac{Lr}{1-(1+r)^{-n}} = \frac{L(\alpha-1)\alpha^n}{\alpha^n-1}, \tag{4.1}$$

where

$$\alpha = 1+r.$$

For instance, if the loan is for \$100,000 to be paid back over 360 months at a nominal yearly interest rate of .09 compounded monthly, then $r = .09/12 = .0075$ and the monthly payment (in dollars) would be

$$A = \frac{100,000(.0075)(1.0075)^{360}}{(1.0075)^{360}-1} = 804.62.$$

Let R_j denote the remaining amount of principal owed after the payment at the end of month j ($j = 0, \ldots, n$). To determine these quantities, note that if one owes R_j at the end of month j then the amount owed immediately before the payment at the end of month $j+1$ is $(1+r)R_j$; because one then pays the amount A, it follows that

$$R_{j+1} = (1+r)R_j - A = \alpha R_j - A.$$

Starting with $R_0 = L$, we obtain:

$$R_1 = \alpha L - A,$$

$$R_2 = \alpha R_1 - A$$
$$= \alpha(\alpha L - A) - A$$
$$= \alpha^2 L - (1 + \alpha)A,$$

$$R_3 = \alpha R_2 - A$$
$$= \alpha(\alpha^2 L - (1 + \alpha)A) - A$$
$$= \alpha^3 L - (1 + \alpha + \alpha^2)A.$$

In general, for $j = 0, \ldots, n$ we obtain

$$R_j = \alpha^j L - A(1 + \alpha + \cdots + \alpha^{j-1})$$
$$= \alpha^j L - A \frac{\alpha^j - 1}{\alpha - 1}$$
$$= \alpha^j L - \frac{L\alpha^n(\alpha^j - 1)}{\alpha^n - 1} \qquad \text{(from (4.1))}$$
$$= \frac{L(\alpha^n - \alpha^j)}{\alpha^n - 1}.$$

Let I_j and P_j denote the amounts of the payment at the end of month j that are for interest and for principal reduction, respectively. Then, since R_{j-1} was owed at the end of the previous month, we have

$$I_j = rR_{j-1}$$
$$= \frac{L(\alpha - 1)(\alpha^n - \alpha^{j-1})}{\alpha^n - 1}$$

and

$$P_j = A - I_j$$
$$= \frac{L(\alpha - 1)}{\alpha^n - 1}[\alpha^n - (\alpha^n - \alpha^{j-1})]$$
$$= \frac{L(\alpha - 1)\alpha^{j-1}}{\alpha^n - 1}.$$

As a check, note that

$$\sum_{j=1}^{n} P_j = L.$$

It follows that the amount of principal repaid in succeeding months increases by the factor $\alpha = 1 + r$. For example, in a \$100,000 loan for 30 years at a nominal interest rate of 9% per year compounded monthly, only \$54.62 of the \$804.62 paid during the first month goes toward reducing the principal of the loan; the remainder is interest. In each succeeding month, the amount of the payment that goes toward the principal increases by the factor 1.0075. □

Consider two cash flow sequences,

$$b_1, b_2, \ldots, b_n \quad \text{and} \quad c_1, c_2, \ldots, c_n.$$

Under what conditions is the present value of the first sequence at least as large as that of the second for every positive interest rate r? Clearly, $b_i \geq c_i$ ($i = 1, \ldots, n$) is a sufficient condition. However, we can obtain weaker sufficient conditions. Let

$$B_i = \sum_{j=1}^{i} b_j \quad \text{and} \quad C_i = \sum_{j=1}^{i} c_j \quad \text{for } i = 1, \ldots, n;$$

then it can be shown that the condition

$$B_i \geq C_i \quad \text{for each } i = 1, \ldots, n$$

suffices. An even weaker sufficient condition is given by the following proposition.

Proposition 4.2.1 *If $B_n \geq C_n$ and if*

$$\sum_{i=1}^{k} B_i \geq \sum_{i=1}^{k} C_i$$

for each $k = 1, \ldots, n$, then

$$\sum_{i=1}^{n} b_i (1 + r)^{-i} \geq \sum_{i=1}^{n} c_i (1 + r)^{-i}$$

for every $r > 0$.

In other words, Proposition 4.2.1 states that the cash flow sequence $b_1, \ldots, b_n$ will, for every positive interest rate r, have a larger present value than the cash flow sequence $c_1, \ldots, c_n$ if (i) the total of the b-cashflows is at least as large as the total of the c-cashflows and (ii) for every $k = 1, \ldots, n$,

$$kb_1 + (k-1)b_2 + \cdots + b_k \geq kc_1 + (k-1)c_2 + \cdots + c_k.$$

4.3 Rate of Return

Consider an investment that, for an initial payment of a $(a > 0)$, returns the amount b after one period. The *rate of return* on this investment is defined to be the interest rate r that makes the present value of the return equal to the initial payment. That is, the rate of return is that value r such that

$$\frac{b}{1+r} = a \quad \text{or} \quad r = \frac{b}{a} - 1.$$

Thus, for example, a $100 investment that returns $150 after one year is said to have a yearly rate of return of .50.

More generally, consider an investment that, for an initial payment of a $(a > 0)$, yields a string of nonnegative returns $b_1, \ldots, b_n$. Here b_i is to be received at the end of period i $(i = 1, \ldots, n)$, and $b_n > 0$. We define the rate of return per period of this investment to be the value of the interest rate such that the present value of the cash flow sequence is equal to zero when values are compounded periodically at that interest rate. That is, if we define the function P by

$$P(r) = -a + \sum_{i=1}^{n} b_i (1+r)^{-i}, \tag{4.2}$$

then the rate of return per period of the investment is that value $r^* > -1$ for which

$$P(r^*) = 0.$$

It follows from the assumptions $a > 0$, $b_i \geq 0$, and $b_n > 0$ that $P(r)$ is a strictly decreasing function of r when $r > -1$, implying (since $\lim_{r \to -1} P(r) = \infty$ and $\lim_{r \to \infty} P(r) = -a < 0$) that there is a unique value r^* satisfying the preceding equation. Moreover, since

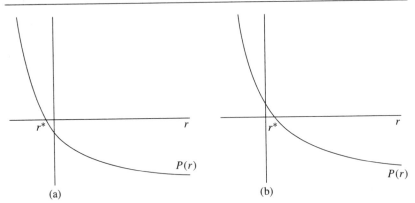

Figure 4.1: $P(r) = -a + \sum_{i \geq 1} b_i (1 + r)^{-i}$: (a) $\sum_i b_i < a$; (b) $\sum_i b_i > a$

$$P(0) = \sum_{i=1}^{n} b_i - a,$$

it follows (see Figure 4.1) that r^* will be positive if

$$\sum_{i=1}^{n} b_i > a$$

and that r^* will be negative if

$$\sum_{i=1}^{n} b_i < a.$$

That is, there is a positive rate of return if the total of the amounts received exceeds the initial investment, and there is a negative rate of return if the reverse holds. Moreover, because of the monotonicity of $P(r)$, it follows that the cash flow sequence will have a positive present value when the interest rate is less than r^* and a negative present value when the interest rate is greater than r^*.

When an investment's rate of return is r^* per period, we often say that the investment yields a $100r^*$-percent rate of return per period.

Example 4.3a   Find the rate of return from an investment that, for an initial payment of 100, yields returns of 60 at the end of each of the first two periods.

Solution. The rate of return will be the solution to

$$100 = \frac{60}{1+r} + \frac{60}{(1+r)^2}.$$

Letting $x = 1/(1+r)$, the preceding can be written as

$$60x^2 + 60x - 100 = 0,$$

which yields that

$$x = \frac{-60 \pm \sqrt{60^2 + 4(60)(100)}}{120}.$$

Since $-1 < r$ implies that $x > 0$, we obtain the solution

$$x = \frac{\sqrt{27,600} - 60}{120} \approx .8844.$$

Hence, the rate of return r^* is such that

$$1 + r^* \approx \frac{1}{.8844} \approx 1.131.$$

That is, the investment yields a rate of return of approximately 13.1% per period. $\qquad\qquad\qquad\qquad\qquad\qquad\qquad\qquad\qquad\qquad$ □

The rate of return of investments whose string of payments spans more than two periods will usually have to be numerically determined. Because of the monotonicity of $P(r)$, a trial-and-error approach is usually quite efficient.

Remarks. (1) If we interpret the cash flow sequence by supposing that $b_1, \ldots, b_n$ represent the successive periodic payments made to a lender who loans a_0 to a borrower, then the lender's periodic rate of return r^* is exactly the effective interest rate per period paid by the borrower.

(2) The quantity r^* is also sometimes called the *internal rate of return*.

Consider now a more general investment cash flow sequence $c_0, c_1, \ldots,$ c_n. Here, if $c_i \geq 0$ then the amount c_i is received by the investor at the end of period i, and if $c_i < 0$ then the amount $-c_i$ must be paid by the investor at the end of period i. If we let

$$P(r) = \sum_{i=0}^{n} c_i(1+r)^{-i}$$

be the present value of this cash flow when the interest rate is r per period, then in general there will not necessarily be a unique solution of the equation

$$P(r) = 0$$

in the region $r > -1$. As a result, the rate-of-return concept is unclear in the case of more general cash flows than the ones considered here. In addition, even in cases where we can show that the preceding equation has a unique solution r^*, it may result that $P(r)$ is not a monotone function of r; consequently, we could *not* assert that the investment yields a positive present value return when the interest rate is on one side of r^* and a negative present value return when it is on the other side.

One general situation for which we can prove that there is a unique solution is when the cash flow sequence starts out negative (resp. positive), eventually becomes positive (negative), and then remains nonnegative (nonpositive) from that point on. In other words, the sequence $c_0, c_1, \ldots, c_n$ has a single sign change. It then follows – upon using Descartes' rule of sign, along with the known existence of at least one solution – that there is a unique solution of the equation $P(r) = 0$ in the region $r > -1$.

4.4   Continuously Varying Interest Rates

Suppose that interest is continuously compounded but with a rate that is changing in time. Let the present time be time 0, and let $r(s)$ denote the interest rate at time s. Thus, if you put x in a bank at time s, then the

amount in your account at time $s + h \approx x(1 + r(s)h)$ (h small).

The quantity $r(s)$ is called the *spot* or the *instantaneous* interest rate at time s.

Let $D(t)$ be the amount that you will have on account at time t if you deposit 1 at time 0. In order to determine $D(t)$ in terms of the interest rates $r(s)$, $0 \le s \le t$, note that (for h small) we have

$$D(s + h) \approx D(s)(1 + r(s)h)$$

or

$$D(s + h) - D(s) \approx D(s)r(s)h$$

or

$$\frac{D(s + h) - D(s)}{h} \approx D(s)r(s).$$

The preceding approximation becomes exact as h becomes smaller and smaller. Hence, taking the limit as $h \to 0$, it follows that

$$D'(s) = D(s)r(s)$$

or

$$\frac{D'(s)}{D(s)} = r(s),$$

implying that

$$\int_0^t \frac{D'(s)}{D(s)} \, ds = \int_0^t r(s) \, ds$$

or

$$\log(D(t)) - \log(D(0)) = \int_0^t r(s) \, ds.$$

Since $D(0) = 1$, we obtain from the preceding equation that

$$D(t) = \exp \left\{ \int_0^t r(s) \, ds \right\}.$$

Now let $P(t)$ denote the present (i.e. time-0) value of the amount 1 that is to be received at time t ($P(t)$ would be the cost of a bond that yields a return of 1 at time t; it would equal e^{-rt} if the interest rate were always equal to r). Because a deposit of $1/D(t)$ at time 0 will be worth 1 at time t, we see that

$$P(t) = \frac{1}{D(t)} = \exp \left\{ -\int_0^t r(s) \, ds \right\}. \tag{4.3}$$

Let $\bar{r}(t)$ denote the average of the spot interest rates up to time t; that is,

$$\bar{r}(t) = \frac{1}{t} \int_0^t r(s) \, ds.$$

The function $\bar{r}(t)$, $t \geq 0$, is called the *yield curve*.

Example 4.4a   Find the yield curve and the present value function if

$$r(s) = \frac{1}{1+s}r_1 + \frac{s}{1+s}r_2.$$

Solution. Rewriting $r(s)$ as

$$r(s) = r_2 + \frac{r_1 - r_2}{1+s}, \quad s \ge 0,$$

shows that the yield curve is given by

$$\bar{r}(t) = \frac{1}{t}\int_0^t \left(r_2 + \frac{r_1 - r_2}{1+s}\right) ds$$

$$= r_2 + \frac{r_1 - r_2}{t}\log(1+t).$$

Consequently, the present value function is

$$P(t) = \exp\{-t\bar{r}(t)\}$$

$$= \exp\{-r_2 t\}\exp\{-\log((1+t)^{r_1-r_2})\}$$

$$= \exp\{-r_2 t\}(1+t)^{r_2-r_1}. \qquad \square$$

4.5   Exercises

Exercise 4.1   What is the effective interest rate when the nominal interest rate of 10% is

(a) compounded semiannually;
(b) compounded quarterly;
(c) compounded continuously?

Exercise 4.2   Suppose that you deposit your money in a bank that pays interest at a nominal rate of 10% per year. How long will it take for your money to double if the interest is compounded continuously?

Exercise 4.3   If you receive 5% interest compounded yearly, approximately how many years will it take for your money to quadruple? What if you were earning only 4%?

Exercise 4.4   Give a formula that approximates the number of years it would take for your funds to triple if you received interest at a rate r compounded yearly.

Exercise 4.5   How much do you need to invest at the beginning of each of the next 60 months in order to have a value of $100,000 at the end of 60 months, given that the annual nominal interest rate will be fixed at 6% and will be compounded monthly?

Exercise 4.6   The yearly cash flows of an investment are

$$-1,000, \ -1,200, \ 800, \ 900, \ 800.$$

Is this a worthwhile investment for someone who can both borrow and save money at the yearly interest rate of 6%?

Exercise 4.7   Consider two possible sequences of end of year returns:

$$20, \ 20, \ 20, \ 15, \ 10, \ 5 \quad \text{and} \quad 10, \ 10, \ 15, \ 20, \ 20, \ 20.$$

Which sequence is preferable if the interest rate, compounded annually, is: (a) 3%; (b) 5%; (c) 10%?

Exercise 4.8   A five-year $10,000 bond with a 10% coupon rate costs $10,000 and pays its holder $500 every six months for five years, with a final additional payment of $10,000 made at the end of those 10 payments. Find its present value if the interest rate is: (a) 6%; (b) 10%; (c) 12%.

Exercise 4.9   A friend purchased a new sound system that was selling for $4,200. He agreed to make a down payment of $1,000 and to make 24 monthly payments of $160, beginning one month from time of purchase. What is the effective interest rate being paid?

Exercise 4.10   Repeat Example 4.2b, this time assuming that the yearly interest rate is 20%.

Exercise 4.11   Repeat Example 4.2b, this time assuming that the cost of a new machine increases by $1,000 each year.

Exercise 4.12   Suppose you have agreed to a bank loan of $120,000, for which the bank charges no fees but 2 points. The quoted interest rate is .5% per month. You are required to pay only the accumulated interest each month for the next 36 months, at which point you must make a balloon payment of the still-owed $120,000. What is the effective interest rate of this loan?

Exercise 4.13   You can pay off a loan either by paying the entire amount of $16,000 now or you can pay $10,000 now and $10,000 at the end of ten years. Which is preferable when the nominal continuously compounded interest rate is: (a) 2%; (b) 5%; (c) 10%?

Exercise 4.14   A U.S. treasury bond (selling at a *par value* of $1,000) that matures at the end of five years is said to have a *coupon rate* of 6% if, after paying $1,000, the purchaser receives $30 at the end of each of the following nine six-month periods and then receives $1,030 at the end of the the tenth period. That is, the bond pays a simple interest rate of 3% per six-month period, with the principal repaid at the end of five years. Assuming a continuously compounded interest rate of 5%, find the present value of such a stream of cash payments.

Exercise 4.15   A zero coupon rate bond having face value F pays the bondholder the amount F when the bond matures. Assuming a continuously compounded interest rate of 8%, find the present value of a zero coupon bond with face value $F = 1,000$ that matures at the end of ten years.

Exercise 4.16   Find the rate of return of a two-year investment that, for an initial payment of 1,000, gives a return at the end of the first year of 500 and a return at the end of the second year of: (a) 300; (b) 500; (c) 700.

Exercise 4.17   Repeat the preceding exercise, reversing the order in which the payments are received.

Exercise 4.18   The inflation rate is defined to be the rate at which prices as a whole are increasing. For instance, if the yearly inflation rate is 4% then what cost $100 last year costs $104 this year. Let r_i denote the inflation rate, and consider an investment whose rate of return is r. We are often interested in determining the investment's rate of return from the point of view of how much the investment increases one's purchasing power; we call this quantity the investment's *inflation-adjusted rate of return* and denote it as r_a. Since the purchasing power of the amount $(1 + r)x$ one year from now is equivalent to that of the amount $(1 + r)x/(1 + r_i)$ today, it follows that – with respect to constant purchasing power units – the investment transforms (in one time period) the amount x into the amount $(1 + r)x/(1 + r_i)$. Consequently, its inflation-adjusted rate of return is

$$r_a = \frac{1 + r}{1 + r_i} - 1.$$

When r and r_i are both small, we have the following approximation:

$$r_a \approx r - r_i.$$

For instance, if a bank pays a simple interest rate of 5% when the inflation rate is 3%, the inflation-adjusted interest rate is approximately 2%. What is its exact value?

Exercise 4.19 Consider an investment cash flow sequence $c_0, c_1, \ldots,$ c_n, where $c_i < 0$, $i < n$, and $c_n > 0$. Show that if

$$P(r) = \sum_{i=0}^{n} c_i (1 + r)^{-i}$$

then, in the region $r > -1$,

(a) there is a unique solution of $P(r) = 0$;
(b) $P(r)$ need not be a monotone function of r.

Exercise 4.20 Suppose you can borrow money at an annual interest rate of 8% but can save money at an annual interest rate of only 5%. If you start with zero capital and if the yearly cash flows of an investment are

$$-1,000, \ 900, \ 800, \ -1,200, \ 700,$$

should you invest?

Exercise 4.21   Show that, if $r(t)$ is an nondecreasing function of t, then so is $\bar{r}(t)$.

Exercise 4.22   Show that the yield curve $\bar{r}(t)$ is a nondecreasing function of t if and only if

$$P(\alpha t) \geq (P(t))^{\alpha} \quad \text{for all } 0 \leq \alpha \leq 1, \; t \geq 0.$$

Exercise 4.23   If $P(t) = e^{-a-bt}$ $(t \geq 0)$, find: (a) $r(t)$; (b) $\bar{r}(t)$.

Exercise 4.24   Show that

$$\text{(a) } r(t) = -\frac{P'(t)}{P(t)} \quad \text{and} \quad \text{(b) } \bar{r}(t) = -\frac{\log P(t)}{t}.$$

Exercise 4.25   Plot the spot interest rate function $r(t)$ of Example 4.4a when

(a) $r_1 < r_2$;
(b) $r_2 < r_1$.

5. Pricing Contracts via Arbitrage

5.1   An Example in Options Pricing

Suppose that the nominal interest rate is r, and consider the following model for pricing an option to purchase a stock at a future time at a fixed price. Let the present price (in dollars) of the stock be 100 per share, and suppose that we know that, after one time period, its price will be either 200 or 50 (see Figure 5.1). Suppose further that, for any y, at a cost of cy you can purchase at time 0 the option to buy y shares of the stock at time 1 at a price of 150 per share. Thus, for instance, if you purchase this option and the stock rises to 200, you would then exercise the option at time 1 and realize a gain of $200 - 150 = 50$ for each of the y options purchased. On the other hand, if the price of the stock at time 1 is 50 then the option would be worthless. In addition to the options, you may also purchase x shares of the stock at time 0 at a cost of $100x$, and each share would be worth either 200 or 50 at time 1.

We will suppose that both x and y can be positive, negative, or zero. That is, you can either buy or sell both the stock and the option. For instance, if x were negative then you would be selling $-x$ shares of stock, yielding you an initial return of $-100x$, and you would then be responsible for buying and returning $-x$ shares of the stock at time 1 at a (time-1) cost of either 200 or 50 per share. (When you sell a stock that you do not own, we say that you are *selling it short*.)

We are interested in determining the appropriate value of c, the unit cost of an option. Specifically, we will show that if r is the one period interest rate then, unless $c = [100 - 50(1+r)^{-1}]/3$, there is a combination of purchases that will always result in a positive present value gain. To show this, suppose that at time 0 we

(a) purchase x units of stock, and
(b) purchase y units of options,

where x and y (both of which can be either positive or negative) are to be determined. The cost of this transaction is $100x + cy$; if this amount

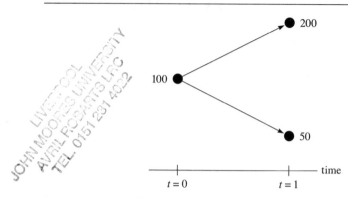

Figure 5.1: Possible Stock Prices at Time 1

is positive, then it should be borrowed from a bank, to be repaid with interest at time 1; if it is negative, then the amount received, $-(100x+cy)$, should be put in the bank to be withdrawn at time 1. The value of our holdings at time 1 depends on the price of the stock at that time and is given by

$$\text{value} = \begin{cases} 200x + 50y & \text{if the price is 200,} \\ 50x & \text{if the price is 50.} \end{cases}$$

This formula follows by noting that, if the stock's price at time 1 is 200, then the x shares of the stock are worth $200x$ and the y units of options to buy the stock at a share price of 150 are worth $(200 - 150)y$. On the other hand, if the stock's price is 50, then the x shares are worth $50x$ and the y units of options are worthless. Now, suppose we choose y so that the value of our holdings at time 1 is the same no matter what the price of the stock at that time. That is, we choose y so that

$$200x + 50y = 50x$$

or

$$y = -3x.$$

Note that y has the opposite sign of x; thus, if $x > 0$ and so x shares of the stock are purchased at time 0, then $3x$ units of stock options are also *sold* at that time. Similarly, if x is negative, then $-x$ shares are sold and $-3x$ units of stock options are purchased at time 0.

Thus, with $y = -3x$, the

$$\text{time-1 value of holdings} = 50x$$

no matter what the value of the stock. As a result, if $y = -3x$ it follows that, after paying off our loan (if $100x + cy > 0$) or withdrawing our money from the bank (if $100x + cy < 0$), we will have gained the amount

$$\begin{aligned} \text{gain} &= 50x - (100x + cy)(1 + r) \\ &= 50x - (100x - 3xc)(1 + r) \\ &= (1 + r)x[3c - 100 + 50(1 + r)^{-1}]. \end{aligned}$$

Thus, if $3c = 100 - 50(1 + r)^{-1}$, then the gain is 0. On the other hand, if $3c \neq 100 - 50(1 + r)^{-1}$, then we can guarantee a positive gain (no matter what the price of the stock at time 1) by letting x be positive when $3c > 100 - 50(1 + r)^{-1}$ and by letting x be negative when $3c < 100 - 50(1 + r)^{-1}$.

For instance, if $(1 + r)^{-1} = .9$ and the cost per option is $c = 20$, then purchasing one share of the stock and selling three units of options initially costs us $100 - 3(20) = 40$, which is borrowed from the bank. However, the value of this holding at time 1 is 50 whether the stock price rises to 200 or falls to 50. Using $40(1 + r) = 44.44$ of this amount to pay our bank loan results in a guaranteed gain of 5.56. Similarly, if the cost of an option is 15, then selling one share of the stock ($x = -1$) and buying three units of options results in an initial gain of $100 - 45 = 55$, which is put into a bank to be worth $55(1 + r) = 61.11$ at time 1. Because the value of our holding at time 1 is -50, a guaranteed profit of 11.11 is attained. A sure-win betting scheme is called an *arbitrage*. Thus, for the numbers considered, the only option cost c that does not result in an arbitrage is $c = (100 - 45)/3 = 55/3$.

Remark. We can also use arbitrage to determine the cost of the option by replicating the option by a combination of borrowing and purchasing the security. To do so, note that if you buy the option at a cost c, then the return from the option at time 1 is

$$\text{return} = \begin{cases} 50 & \text{if the stock's price is 200,} \\ 0 & \text{if the stock's price is 50.} \end{cases}$$

Now, suppose that at time 0 you borrow from a bank the amount

$$\frac{50}{1+r}y$$

and put up from your own funds the amount

$$\left(100 - \frac{50}{1+r}\right)y$$

to purchase y shares of the stock. At time 1 you will owe the bank $50y$; so, after selling the stock and paying off this debt, you will have a return given by

$$\text{return} = \begin{cases} 150y & \text{if the stock's price is 200,} \\ 0 & \text{if the stock's price is 50.} \end{cases}$$

Therefore, when $y = 1/3$, the return at time 1 is the same as that from the option. In other words, by paying the initial amount $\left(100 - \frac{50}{1+r}\right)/3$ (and borrowing the remaining amount needed to purchase $1/3$ shares of stock) you can replicate the payoff of the option. These two investment returns are identical, so their costs must be also if one is to avoid an arbitrage. Thus, if $c > \left(100 - \frac{50}{1+r}\right)/3$, then a sure profit can be made by using your own funds along with borrowed money to purchase $1/3$ shares of stock while simultaneously selling one option share. If $c < \left(100 - \frac{50}{1+r}\right)/3$, then a sure profit can be made by selling $1/3$ shares of the stock, using c of the returns of this sale to purchase the option, and depositing the remaining $100/3 - c$ in a bank.

5.2   Other Examples of Pricing via Arbitrage

The type of option considered in Section 5.1 is known as a *call* option because it gives one the option of calling for the stock at a specified price, known as the *exercise* or *strike* price. An *American style* call option allows the buyer to exercise the option at any time up to the expiration time, whereas a *European style* call option can only be exercised at the expiration time. Although it might seem that, because of its additional flexibility, the American style option should be worth more, it turns out that it is never optimal to exercise a call option early; thus, the two style options have identical worths. We now prove this claim.

Proposition 5.2.1 *One should never exercise an American style call option before its expiration time t.*

Proof. Suppose that the present price of the stock is S, that you own an option to buy one share of the stock at a fixed price K, and that the option expires after an additional time t. As soon as you exercise the option, you will realize the amount $S - K$. However, consider what would transpire if, instead of exercising the option, you sell the stock short and then purchase the stock at the exercise time, either by paying the market price at that time or by exercising your option and paying K, whichever is less expensive. Under this strategy, you will initially receive S, and will then have to pay the minimum of the market price and the exercise price K after an additional time t. This is clearly preferable to receiving S and immediately paying out K.   □

Our next example uses arbitrage considerations to derive a lower bound on the cost of a call option.

Example 5.2a Let C denote the cost of an option to purchase a security at time t for the strike price K. If S is the present price of the security and r is the continuously compounded interest rate, then

$$C \geq S - Ke^{-rt}.$$

To see why this must be true, suppose to the contrary that

$$C < S - Ke^{-rt}$$

or (equivalently) that

$$(S - C)e^{rt} > K.$$

However, if the foregoing inequality held then we could guarantee a sure win by

(1) buying the option and
(2) selling the security.

These two transactions will lead to a cash gain of $S - C$, which we can use to purchase a bond that matures at time t. At time t, we then return the security previously sold by buying it for a cost equal to the minimum of the strike price K of the option we own or the market price at time t.

Consequently, at time t we will receive $(S - C)e^{rt}$ from the bond, and then use at most K of this amount to purchase the security, thus yielding us a positive gain of at least

$$(S - C)e^{rt} - K. \qquad \square$$

In addition to call options there are also *put* options on stocks. These give their owners the option of putting a stock up for sale at a specified price. An American style put option allows the owner to put the stock up for sale – that is, to exercise the option – at any time up to the expiration time of the option. A European style put option can only be exercised at its expiration time. Contrary to the situation with call options, it may be advantageous to exercise a put option before its expiration time, and so the American style put option may be worth more than the European. The absence of arbitrage implies a relationship between the price of a European put option having exercise price K and expiration time t and the price of a call option on that stock that also has exercise price K and expiration time t. This is known as the *put–call option parity formula,* which may be stated as follows.

Proposition 5.2.2 *Let C be the price of a call option that enables its holder to buy one share of a stock at an exercise price K at time t; also, let P be the price of a European put option that enables its holder to sell one share of the stock for the amount K at time t. Let S be the price of the stock at time* 0. *Then, assuming that interest is continuously discounted at a nominal rate r, either*

$$S + P - C = Ke^{-rt}$$

or there is an arbitrage opportunity.

Proof. If

$$S + P - C < Ke^{-rt}$$

then we can effect a sure win by initially buying one share of the stock, buying one put option, and selling one call option. This initial payout of $S + P - C$ is borrowed from a bank to be repaid at time t. Let us now consider the value of our holdings at time t. There are two cases that depend on $S(t)$, the stock's market price at time t. If $S(t) \leq K$ then the call option we sold is worthless, and we can exercise our put option to sell the stock for the amount K. On the other hand, if $S(t) > K$ then our put option is worthless and the call option we sold will be exercised, forcing us to sell our stock for the price K. Thus, in either case we will realize the amount K at time t. Since $K > e^{rt}(S + P - C)$, we can pay off our bank loan and realize a positive profit in all cases.

When

$$S + P - C > Ke^{-rt},$$

we can make a sure profit by reversing the procedure just described. Namely, we now sell one share of stock, sell one put option, and buy one call option. We leave the details of the verification to the reader. □

The arbitrage principle also determines the relationship between the present price of a stock and the contracted price to buy the stock at a specified time in the future. Our next two examples are related to these *forwards contracts*.

Example 5.2b *Forwards Contracts* Let S be the present market price of a specified stock. In a forwards agreement, one agrees at time 0 to pay the amount F at time t for one share of the stock that will be delivered at the time of payment. That is, one contracts a price for the stock, which is to be delivered and paid for at time t. We will now present an arbitrage argument to show that, if interest is continuously discounted at the nominal interest rate r, then in order for there to be no arbitrage opportunity we must have

$$F = Se^{rt}.$$

To see why this equality must hold, suppose that instead

$$F < Se^{rt}.$$

In this case, a sure win is obtained by selling the stock at time 0 with the understanding that you will buy it back at time t. Put the sale proceeds S into a bond that matures at time t and, in addition, buy a forwards contract for delivery of one share of the stock at time t. Thus, at time t you will receive Se^{rt} from your bond. From this, you pay F to obtain one share of the stock, which you then return to settle your obligation. You thus end with a positive profit of $Se^{rt} - F$. On the other hand, if

$$F > Se^{rt}$$

then you can guarantee a profit of $F - Se^{rt}$ by simultaneously selling a forwards contract and borrowing S to purchase the stock. At time t you will receive F for your stock, out of which you repay your loan amount of Se^{rt}.   □

When one purchases a share of a stock in the stock market, one is purchasing a share of ownership in the entity that issues the stock. On the other hand, the commodity market deals with more concrete objects: agricultural items like oats, corn, or wheat; energy products like crude oil and natural gas; metals such as gold, silver, or platinum; animal parts such as hogs, pork-bellies, and beef; and so on. Almost all of the activity on the commodities market is involved with contracts for future purchases and sales of the commodity. Thus, for instance, you could purchase a contract to buy natural gas in 90 days for a price that is specified today. (Such a *futures contract* differs from a forwards contract in that, although one pays in full when delivery is taken for both, in futures contracts one settles up on a daily basis depending on the change of the price of the futures contract on the commodity exchange.) You could also write a futures contract that obligates you to sell gas at a specified price at a specified time. Most people who play the commodities market never have any actual contact with the commodity. Rather, an individual who buys a futures contract most often sells that contract before the delivery date.

The relationship given in Example 5.2b does *not* hold for futures contracts in the commodity market. For one thing, if $F > Se^{rt}$ and you purchase the commodity (say, crude oil) to sell back at time t, then you will incur additional costs related to storing and insuring the oil. Also, when $F < Se^{rt}$, selling the commodity for today's price requires that you be able to deliver it immediately.

One of the most popular types of forward contracts involves currency exchanges, the topic of our next example.

Example 5.2c   The September 4, 1998, edition of the *New York Times* gives the following listing for the price of a German mark (or DM):

- today – .5777;
- 90-day forward – .5808.

In other words, you can purchase 1 DM today at the price of $.5777. In addition, you can sign a contract to purchase 1 DM in 90 days at a price, to be paid on delivery, of $.5808. Why are these prices different?

Solution. One might suppose that the difference is caused by the market's expectation of the worth in 90 days of the German DM relative to the U.S. dollar. However, it turns out that the entire price differential is due to the different interest rates in Germany and in the United States. Suppose that interest in both countries is continuously compounded at nominal yearly rates: r_u in the United States and r_g in Germany. Let S denote the present price of 1 DM, and let F be the price for a forwards contract to be delivered at time t. (This example considers the special case where $S = .5777$, $F = .5808$, and $t = 90/365$.) We now argue that, in order for there not to be an arbitrage opportunity, we must have

$$F = Se^{(r_u - r_g)t}.$$

To see why, suppose first that

$$Fe^{r_g t} > Se^{r_u t}.$$

To obtain a sure win, borrow S dollars to be repaid at time t. Use these dollars to buy 1 DM, which in turn is used to buy a German bond that matures at time t. (Thus, at time t you will have $e^{r_g t}$ German marks and you will owe an American bank $Se^{r_u t}$ dollars.) Also, write a contract to sell $e^{r_g t}$ German marks at time t for a total price of $Fe^{r_g t}$ dollars. Then, at time t you sell your German marks for $Fe^{r_g t}$ dollars, use $Se^{r_u t}$ of this to pay off your American debt, and end with a profit of $Fe^{r_g t} - Se^{r_u t}$.
   On the other hand, if

$$Fe^{r_g t} < Se^{r_u t}$$

then you can obtain a sure win by reversing the preceding operation as follows: At time 0, buy a futures contract to purchase $e^{r_g t}$ DM at time t; borrow 1 DM from a German bank and sell it for S dollars, which you then use to buy an American bond maturing at time t. Thus, at time t you will have $Se^{r_u t}$ dollars; use $Fe^{r_g t}$ of it to pay for your futures contract. This gives you $e^{r_g t}$ marks, which you then use to retire your German bank loan debt. Hence, you end with a positive profit of $Se^{r_u t} - Fe^{r_g t}$. □

5.3   Exercises

Exercise 5.1   For the example of Section 5.1, go through the details to show that if $c < \left(100 - \frac{50}{1+r}\right)/3$ then a sure win is obtained by using the strategy given in the Remark at the end of that section.

Exercise 5.2   Suppose it is known that the price of a certain security after one period will be one of the r values $s_1, \ldots, s_r$. What should be the cost of an option to purchase the security at time 1 for the price K when $K < \min s_i$?

Exercise 5.3   Let C be the price of a call option to purchase a security whose present price is S. Argue that $C \le S$.

Exercise 5.4   Let P be the price of a put option to sell a security, whose present price is S, for the amount K. Which of the following are true?

(a) $P \le S$;
(b) $P \le K$.

Exercise 5.5   Let P be the price of a put option to sell a security, whose present price is S, for the amount K. Argue that

$$P \ge Ke^{-rt} - S,$$

where t is the exercise time and r is the interest rate.

Exercise 5.6   With regard to Proposition 5.2.2, verify that the strategy of selling one share of stock, selling one put option, and buying one call option is always a winning strategy when $S + P - C > Ke^{-rt}$.

Exercise 5.7   Explain why the price of an American put option having exercise time t cannot be less than the price of a second put option on the same security that is identical to the first option except that its exercise time is earlier.

Exercise 5.8   Is the result of the preceding exercise valid for European puts?

Exercise 5.9 If a stock is selling for a price s immediately before it pays a dividend d (i.e., the amount d per share is paid to every shareholder), then what should its price be immediately after the dividend is paid?

Exercise 5.10 Let $S(t)$ be the price of a given security at time t. All of the following options have exercise time t and (unless stated otherwise) exercise price K. Give the payoff at time t that is earned by an investor who:

(a) owns one call and one put option;
(b) owns one call having exercise price K_1 and has sold one put having exercise price K_2;
(c) owns two calls and has sold short one share of the security;
(d) owns one share of the security and has sold one call.

REFERENCES

[1] Cox, J., and M. Rubinstein (1985). *Options Markets.* Englewood Cliffs, NJ: Prentice-Hall.
[2] Merton, R. (1973). "Theory of Rational Option Pricing." *Bell Journal of Economics and Mangagement Science* 4: 141–83.
[3] Samuelson, P., and R. Merton (1969). "A Complete Model of Warrant Pricing that Maximizes Utility." *Industrial Management Review* 10: 17–46.
[4] Stoll, H. R., and R. E. Whaley (1986). "New Option Intruments: Arbitrageable Linkages and Valuation." *Advances in Futures and Options Research* 1 (part A): 25–62.

6. The Arbitrage Theorem

6.1   The Arbitrage Theorem

Consider an experiment whose set of possible outcomes is $\{1, 2, \ldots, m\}$, and suppose that n wagers concerning this experiment are available. If the amount x is bet on wager i, then $xr_i(j)$ is received if the outcome of the experiment is j ($j = 1, \ldots, m$). In other words, $r_i(\cdot)$ is the *return function* for a unit bet on wager i. The amount bet on a wager is allowed to be positive, negative, or zero.

A betting strategy is a vector $\mathbf{x} = (x_1, x_2, \ldots, x_n)$, with the interpretation that x_1 is bet on wager 1, x_2 is bet on wager 2, $\ldots$, x_n is bet on wager n. If the outcome of the experiment is j, then the return from the betting strategy $\mathbf{x}$ is given by

$$\text{return from } \mathbf{x} = \sum_{i=1}^{n} x_i r_i(j).$$

The following result, known as the *arbitrage theorem,* states that either there exists a probability vector $\mathbf{p} = (p_1, p_2, \ldots, p_m)$ on the set of possible outcomes of the experiment under which the expected return of each wager is equal to zero, or else there exists a betting strategy that yields a positive win for each outcome of the experiment.

Theorem 6.1.1 (The Arbitrage Theorem)   *Exactly one of the following is true: Either*

(a) *there is a probability vector* $\mathbf{p} = (p_1, p_2, \ldots, p_m)$ *for which*

$$\sum_{j=1}^{m} p_j r_i(j) = 0 \quad \text{for all } i = 1, \ldots, n,$$

or else

(b) *there is a betting strategy* $\mathbf{x} = (x_1, x_2, \ldots, x_n)$ *for which*

$$\sum_{i=1}^{n} x_i r_i(j) > 0 \quad \text{for all } j = 1, \ldots, m.$$

Proof. See Section 6.3.

If X is the outcome of the experiment, then the arbitrage theorem states that either there is a set of probabilities $(p_1, p_2, \ldots, p_m)$ such that if

$$P\{X = j\} = p_j \quad \text{for all } j = 1, \ldots, m$$

then

$$E[r_i(X)] = 0 \quad \text{for all } i = 1, \ldots, n,$$

or else there is a betting strategy that leads to a sure win. In other words, either there is a probability vector on the outcomes of the experiment that results in all bets being fair, or else there is a betting scheme that guarantees a win.

Example 6.1a In some situations, the only type of wagers allowed are ones that choose one of the outcomes i $(i = 1, \ldots, m)$ and then bet that i is the outcome of the experiment. The return from such a bet is often quoted in terms of *odds*. If the odds against outcome i are o_i (often expressed as "o_i to 1"), then a one-unit bet will return either o_i if i is the outcome of the experiment or -1 if i is not the outcome. That is, a one-unit bet on i will either win o_i or lose 1. The return function for such a bet is given by

$$r_i(j) = \begin{cases} o_i & \text{if } j = i, \\ -1 & \text{if } j \neq i. \end{cases}$$

Suppose that the odds $o_1, o_2, \ldots, o_m$ are quoted. In order for there not to be a sure win, there must be a probability vector $\mathbf{p} = (p_1, p_2, \ldots, p_m)$ such that, for each i $(i = 1, \ldots, m)$,

$$0 = E_\mathbf{p}[r_i(X)] = o_i p_i - (1 - p_i).$$

That is, we must have

$$p_i = \frac{1}{1 + o_i}.$$

Since the p_i must sum to 1, this means that the condition for there not to be an arbitrage is that

$$\sum_{i=1}^{m} \frac{1}{1+o_i} = 1.$$

That is, if $\sum_{i=1}^{m}(1+o_i)^{-1} \neq 1$, then a sure win is possible. For instance, suppose there are three possible outcomes and the quoted odds are as follows.

Outcome	Odds
1	1
2	2
3	3

That is, the odds against outcome 1 are 1 to 1; they are 2 to 1 against outcome 2; and they are 3 to 1 against outcome 3. Since

$$\frac{1}{2} + \frac{1}{3} + \frac{1}{4} = \frac{13}{12} \neq 1,$$

a sure win is possible. One possibility is to bet -1 on outcome 1 (so you either win 1 if the outcome is not 1 or you lose 1 if the outcome is 1) and bet $-.7$ on outcome 2 (so you either win .7 if the outcome is not 2 or you lose 1.4 if it is 2), and $-.5$ on outcome 3 (so you either win .5 if the outcome is not 3 or you lose 1.5 if it is 3). If the experiment results in outcome 1, you win $-1 + .7 + .5 = .2$; if it results in outcome 2, you win $1 - 1.4 + .5 = .1$; if it results in outcome 3, you win $1 + .7 - 1.5 = .2$. Hence, in all cases you win a positive amount. □

Example 6.1b Let us reconsider the option pricing example of Section 5.1, where the initial price of a stock is 100 and the price after one period is assumed to be either 200 or 50. At a cost of c per share, we can purchase at time 0 the option to buy the stock at time 1 for the price of 150. For what value of c is no sure win possible?

Solution. In the context of this section, the outcome of the experiment is the value of the stock at time 1; thus, there are two possible outcomes.

There are also two different wagers: to buy (or sell) the stock, and to buy (or sell) the option. By the arbitrage theorem, there will be no sure win if there are probabilities $(p, 1 - p)$ on the outcomes that make the expected present value return equal to zero for both wagers.

The present value return from purchasing one share of the stock is

$$\text{return} = \begin{cases} 200(1 + r)^{-1} - 100 & \text{if the price is 200 at time 1,} \\ 50(1 + r)^{-1} - 100 & \text{if the price is 50 at time 1.} \end{cases}$$

Hence, if p is the probability that the price is 200 at time 1, then

$$E[\text{return}] = p\left[\frac{200}{1 + r} - 100\right] + (1 - p)\left[\frac{50}{1 + r} - 100\right]$$

$$= p\frac{150}{1 + r} + \frac{50}{1 + r} - 100.$$

Setting this equal to zero yields that

$$p = \frac{1 + 2r}{3}.$$

Therefore, the only probability vector $(p, 1 - p)$ that results in a zero expected return for the wager of purchasing the stock has $p = (1 + 2r)/3$.

In addition, the present value return from purchasing one option is

$$\text{return} = \begin{cases} 50(1 + r)^{-1} - c & \text{if the price is 200 at time 1,} \\ -c & \text{if the price is 50 at time 1.} \end{cases}$$

Hence, when $p = (1 + 2r)/3$, the expected return of purchasing one option is

$$E[\text{return}] = \frac{1 + 2r}{3}\frac{50}{1 + r} - c.$$

It thus follows from the arbitrage theorem that the only value of c for which there will not be a sure win is

$$c = \frac{1 + 2r}{3}\frac{50}{1 + r};$$

that is, when

$$c = \frac{50 + 100r}{3(1 + r)},$$

which is in accord with the result of Section 5.1.   □

6.2   The Multiperiod Binomial Model

Let us now consider a stock option scenario in which there are n periods and where the nominal interest rate is r per period. Let $S(0)$ be the initial price of the stock, and for $i = 1, \ldots, n$ let $S(i)$ be its price at i time periods later. Suppose that $S(i)$ is either $uS(i-1)$ or $dS(i-1)$, where $d < 1+r < u$. That is, going from one time period to the next, the price either goes up by the factor u or down by the factor d. Furthermore, suppose that at time 0 an option may be purchased that enables one to buy the stock after n periods have passed for the amount K. In addition, the stock may be purchased and sold anytime within these n time periods.

Let X_i equal 1 if the stock's price goes up by the factor u from period $i-1$ to i, and let it equal 0 if that price goes down by the factor d. That is,

$$X_i = \begin{cases} 1 & \text{if } S(i) = uS(i-1), \\ 0 & \text{if } S(i) = dS(i-1). \end{cases}$$

The outcome of the experiment can now be regarded as the value of the vector $(X_1, X_2, \ldots, X_n)$. It follows from the arbitrage theorem that, in order for there not to be an arbitrage opportunity, there must be probabilities on these outcomes that make all bets fair. That is, there must be a set of probabilities

$$P\{X_1 = x_1, \ldots, X_n = x_n\}, \quad x_i = 0, 1, \ i = 1, \ldots, n,$$

that make all bets fair.

Now consider the following type of bet: First choose a value of i ($i = 1, \ldots, n$) and a vector $(x_1, \ldots, x_{i-1})$ of zeros and ones, and then observe the first $i-1$ changes. If $X_j = x_j$ for each $j = 1, \ldots, i-1$, immediately buy one unit of stock and then sell it back the next period. If the stock is purchased, then its cost at time $i-1$ is $S(i-1)$; the time-$(i-1)$ value of the amount obtained when it is then sold at time i is either $(1+r)^{-1}uS(i-1)$ if the stock goes up or $(1+r)^{-1}dS(i-1)$ if it goes down. Therefore, if we let

$$\alpha = P\{X_1 = x_1, \ldots, X_{i-1} = x_{i-1}\}$$

denote the probability that the stock is purchased, and let

$$p = P\{X_i = 1 \mid X_1 = x_1, \ldots, X_{i-1} = x_{i-1}\}$$

denote the probability that a purchased stock goes up the next period, then the expected gain on this bet (in time-$(i-1)$ units) is

$$\alpha[p(1+r)^{-1}uS(i-1) + (1-p)(1+r)^{-1}dS(i-1) - S(i-1)].$$

Consequently, the expected gain on this bet will be zero, provided that

$$\frac{pu}{1+r} + \frac{(1-p)d}{1+r} = 1$$

or, equivalently, that

$$p = \frac{1+r-d}{u-d}.$$

In other words, the only probability vector that results in an expected gain of zero for this type of bet has

$$P\{X_i = 1 \mid X_1 = x_1, \ldots, X_{i-1} = x_{i-1}\} = \frac{1+r-d}{u-d}.$$

Since $x_1, \ldots, x_n$ are arbitrary, this implies that the only probability vector on the set of outcomes that results in all these bets being fair is the one that takes $X_1, \ldots, X_n$ to be independent random variables with

$$P\{X_i = 1\} = p = 1 - P\{X_i = 0\}, \quad i = 1, \ldots, n, \qquad (6.1)$$

where

$$p = \frac{1+r-d}{u-d}. \qquad (6.2)$$

It can be shown that, with these probabilities, any bet on buying stock will have zero expected gain. Thus, it follows from the arbitrage theorem that either the cost of the option must be equal to the expectation of the present (i.e., the time-0) value of owning it using the preceding probabilities, or else there will be an arbitrage opportunity. So, to determine the no-arbitrage cost, assume that the X_i are independent 0-or-1 random variables whose common probability p of being equal to 1 is given by Equation (6.2). Letting Y denote their sum, it follows that Y is just the number of the X_i that are equal to 1, and thus Y is a binomial random variable with parameters n and p. Now, in going from period to period, the stock's price is its old price multiplied by either u or by d. At time n, the price would have gone up Y times and down $n - Y$ times, so it follows that the stock's price after n periods can be expressed as

$$S(n) = u^Y d^{n-Y} S(0),$$

where $Y = \sum_{i=1}^{n} X_i$ is, as previously noted, a binomial random variable with parameters n and p. The value of owning the option after n periods have elapsed is $(S_n - K)^+$, which is defined to equal either $S_n - K$ (when this quantity is nonnegative) or zero (when it is negative). Therefore, the present (time-0) value of owning the option is

$$(1+r)^{-n}(S(n) - K)^+$$

and so the expectation of the present value of owning the option is

$$(1+r)^{-n} E[(S(n) - K)^+] = (1+r)^{-n} E[(S(0)u^Y d^{n-Y} - K)^+].$$

Thus, the only option cost C that does not result in an arbitrage is

$$C = (1+r)^{-n} E[(S(0)u^Y d^{n-Y} - K)^+]. \qquad (6.3)$$

Remark. Although Equation (6.3) could be streamlined for computational convenience, the expression as given is sufficient for our main purpose: determining the unique no-arbitrage option cost when the underlying security follows a geometric Brownian motion. This is accomplished in our next chapter, where we derive the famous Black–Scholes formula.

6.3   Proof of the Arbitrage Theorem

In order to prove the arbitrage theorem, we first present the duality theorem of linear programming as follows. Suppose that, for given constants c_i, b_j, and $a_{i,j}$ ($i = 1, \ldots, n$, $j = 1, \ldots, m$), we want to choose values $x_1, \ldots, x_n$ that will

$$\text{maximize} \quad \sum_{i=1}^{n} c_i x_i$$

$$\textit{subject to}$$

$$\sum_{i=1}^{n} a_{i,j} x_i \le b_j, \quad j = 1, 2, \ldots, m.$$

This problem is called a primal linear program. Every primal linear program has a *dual* problem, and the dual of the preceding linear program is to choose values $y_1, \ldots, y_m$ that

$$\text{minimize} \sum_{j=1}^{m} b_j y_j$$

subject to

$$\sum_{j=1}^{m} a_{i,j} y_j = c_i, \quad i = 1, \ldots, n,$$

$$y_j \geq 0, \quad j = 1, \ldots, m.$$

A linear program is said to be *feasible* if there are variables $(x_1, \ldots, x_n$ in the primal linear program or $y_1, \ldots, y_m$ in the dual) that satisfy the constraints. The key theoretical result of linear programming is the *duality theorem*, which we state without proof.

Proposition 6.3.1 (Duality Theorem of Linear Programming) *If a primal and its dual linear program are both feasible, then they both have optimal solutions and the maximal value of the primal is equal to the minimal value of the dual. If either problem is infeasible, then the other does not have an optimal solution.*

A consequence of the duality theorem is the arbitrage theorem. Recall that the arbitrage theorem refers to a situation in which there are n wagers with payoffs that are determined by the result of an experiment having possible outcomes $1, 2, \ldots, m$. Specifically, if you bet wager i at level x, then you win the amount $x r_i(j)$ if the outcome of the experiment is j. A betting strategy is a vector $\mathbf{x} = (x_1, \ldots, x_n)$, where each x_i can be positive or negative (or zero), and with the interpretation that you simultaneously bet wager i at level x_i for each $i = 1, \ldots, n$. If the outcome of the experiment is j, then your winnings from the betting strategy $\mathbf{x}$ are

$$\sum_{i=1}^{n} x_i r_i(j).$$

Proposition 6.3.2 (Arbitrage Theorem)   *Exactly one of the following is true: Either*

(i) *there exists a probability vector* $\mathbf{p} = (p_1, \ldots, p_m)$ *for which*

$$\sum_{j=1}^{m} p_j r_i(j) = 0 \quad \text{for all } i = 1, \ldots, n;$$

or

(ii) *there exists a betting strategy* $\mathbf{x} = (x_1, \ldots, x_n)$ *such that*

$$\sum_{i=1}^{n} x_i r_i(j) > 0 \quad \text{for all } j = 1, \ldots, m.$$

That is, either there exists a probability vector under which all wagers have expected gain equal to zero, or else there is a betting strategy that always results in a positive win.

Proof. Let x_{n+1} denote an amount that the gambler can be sure of winning, and consider the problem of maximizing this amount. If the gambler uses the betting strategy $(x_1, \ldots, x_n)$ then she will win $\sum_{i=1}^{n} x_i r_i(j)$ if the outcome of the experiment is j. Hence, she will want to choose her betting strategy $(x_1, \ldots, x_n)$ and x_{n+1} so as to

$$\text{maximize } x_{n+1}$$

$$\text{subject to}$$

$$\sum_{i=1}^{n} x_i r_i(j) \geq x_{n+1}, \quad j = 1, \ldots, m.$$

Letting

$$a_{i,j} = -r_i(j), \quad i = 1, \ldots, n, \quad a_{n+1,j} = 1,$$

we can rewrite the preceding as follows:

$$\text{maximize } x_{n+1}$$

$$\text{subject to}$$

$$\sum_{i=1}^{n+1} a_{i,j} x_i \leq 0, \quad j = 1, \ldots, m.$$

Note that the preceding linear program has $c_1 = c_2 = \cdots = c_n = 0$, $c_{n+1} = 1$, and upper-bound constraint values all equal to zero (i.e., all $b_j = 0$). Consequently, its dual program is to choose variables $y_1, \ldots, y_m$ so as to

$$\text{minimize } 0$$

$$\text{subject to}$$

$$\sum_{j=1}^{m} a_{i,j} y_j = 0, \quad i = 1, \ldots, n,$$

$$\sum_{j=1}^{m} a_{n+1,j} y_j = 1,$$

$$y_j \geq 0, \quad j = 1, \ldots, m.$$

Using the definitions of the quantities $a_{i,j}$ gives that this dual linear program can be written as

$$\text{minimize } 0$$

$$\textit{subject to}$$

$$\sum_{j=1}^{m} r_i(j) y_j = 0, \quad i = 1, \ldots, n,$$

$$\sum_{j=1}^{m} y_j = 1,$$

$$y_j \geq 0, \quad j = 1, \ldots, m.$$

Observe that this dual will be feasible, and its minimal value will be zero, if and only if there is a probability vector $(y_1, \ldots, y_m)$ under which all wagers have expected return 0. The primal problem is feasible because $x_i = 0$ $(i = 1, \ldots, n + 1)$ satisfies its constraints, so it follows from the duality theorem that if the dual problem is also feasible then the optimal value of the primal is zero and hence no sure win is possible. On the other hand, if the dual is infeasible then it follows from the duality theorem that there is no optimal solution of the primal. But this implies that zero is not the optimal solution, and thus there is a betting scheme whose minimal return is positive. (The reason there is no primal optimal solution when the dual is infeasible is because the primal is unbounded in this case. That is, if there is a betting scheme $\mathbf{x}$ that gives a guaranteed return of at least $v > 0$, then $c\mathbf{x}$ gives a guaranteed return of at least cv.) □

6.4 Exercises

Exercise 6.1 Consider an experiment with three possible outcomes and odds as follows.

Outcome	Odds
1	1
2	2
3	5

Is there a betting scheme that results in a sure win?

Exercise 6.2   Consider an experiment with four possible outcomes, and suppose that the quoted odds for the first three of these outcomes are as follows.

Outcome	Odds
1	2
2	3
3	4

What must be the odds against outcome 4 if there is to be no possible arbitrage when one is allowed to bet both for and against any of the outcomes?

Exercise 6.3   Repeat Exercise 6.1 when the odds are as follows.

Outcome	Odds
1	2
2	2
3	2

Exercise 6.4   Suppose, in Exercise 6.1, that one is also allowed to choose any pair of outcomes $i \neq j$ and bet that the outcome will be either i or j. What should the odds be on these three bets if an arbitrage opportunity is to be avoided?

Exercise 6.5   In Example 6.1a, show that if

$$\sum_{i=1}^{m} \frac{1}{1+o_i} \neq 1$$

then the betting scheme

$$x_i = \frac{(1+o_i)^{-1}}{1 - \sum_{i=1}^{m}(1-o_i)^{-1}}, \quad i = 1, \ldots, m,$$

will always yield a gain of exactly 1.

Exercise 6.6   In Example 6.1b, suppose one also has the option of purchasing a put option that allows its holder to put the stock for sale at the end of one period for a price of 150. Determine the value of P, the cost of the put, if there is to be no arbitrage; then show that the resulting call and put prices satisfy the put–call option parity formula (Proposition 5.2.2).

Exercise 6.7   Suppose that, in each period, the cost of a security either goes up by a factor of 2 or down by a factor of $1/2$ (i.e., $u = 2$, $d = 1/2$). If the initial price of the security is 100, determine the no-arbitrage cost of a call option to purchase the security at the end of two periods for a price of 150.

Exercise 6.8   Suppose, in Example 6.1b, that there are three possible prices for the security at time 1: 50, 100, or 200. (That is, allow for the possibility that the security's price remains unchanged.) Use the arbitrage theorem to find an interval for which there is no arbitrage if c lies in that interval.

A betting strategy $\mathbf{x}$ such that (using the notation of Section 6.1)

$$\sum_{i=1}^{n} x_i r_i(j) \geq 0, \quad i = 1, \ldots, m,$$

with strict inequality for at least one i, is said to be a *weak arbitrage* strategy. That is, whereas an arbitrage is present if there is a strategy that results in a positive gain for every outcome, a weak arbitrage is present if there is a strategy that never results in a loss and results in a positive gain for at least one outcome. (An arbitrage can be thought of as a *free lunch*, whereas a weak arbitrage is a *free lottery ticket*.) It can be shown that there will be no weak arbitrage if and only if there is a probability vector $\mathbf{p}$, all of whose components are positive, such that

$$\sum_{j=1}^{m} p_j r_i(j) = 0, \quad i = 1, \ldots, n.$$

In other words, there will be no weak arbitrage if there is a probability vector that gives positive weight to each possible outcome and makes all bets fair.

Exercise 6.9 In Exercise 6.8, show that a weak arbitrage is possible if the cost of the option is equal to either endpoint of the interval determined.

Exercise 6.10 For the model of Section 6.2 with $n = 1$, show how an option can be replicated by a combination of borrowing and buying the security.

REFERENCES

[1] De Finetti, Bruno (1937). "La prevision: ses lois logiques, ses sources subjectives." *Annales de l'Institut Henri Poincaré* 7: 1–68; English translation in S. Kyburg (Ed.) (1962), *Studies in Subjective Probability,* pp. 93–158. New York: Wiley.
[2] Gale, David (1960). *The Theory of Linear Economic Models.* New York: McGraw-Hill.

7. The Black–Scholes Formula

7.1   The Black–Scholes Formula

Consider a call option having strike price K and exercise time t. That is, the option allows one to purchase a single unit of an underlying security at time t for the price K. Suppose further that the nominal interest rate is r, compounded continuously, and also that the price of the security follows a geometric Brownian motion with variance parameter σ^2. Under these assumptions, we will find the unique cost of the option that does not give rise to an arbitrage.

To begin, recall from Section 3.2 that the first t time units of a geometric Brownian motion with variance parameter σ^2 can be approximated by a process that, at each time point $t/n, 2t/n, \ldots, nt/n$, either goes up by the factor

$$u = e^{\sigma\sqrt{t/n}} \approx 1 + \sigma\sqrt{t/n} + \frac{\sigma^2 t}{2n} \tag{7.1}$$

or down by the factor

$$d = e^{-\sigma\sqrt{t/n}} \approx 1 - \sigma\sqrt{t/n} + \frac{\sigma^2 t}{2n}, \tag{7.2}$$

where n is a large positive integer (and where the approximations of u and d are obtained by taking the first three terms of the Taylor series expansion about 0 of the function e^x). As a consequence, we see that the first t time units of *every* geometric Brownian motion having variance parameter σ^2, no matter what the value of its other parameter μ, can be approximated by an n-period binomial model whose up and down factors are given by Equations (7.1) and (7.2). But we know from Section 6.2 that there is a unique no-arbitrage option cost C for this approximation model. Because the n-period approximation model to geometric Brownian motion becomes exact as n increases, it follows that, with rt/n as the one-period interest rate, C will converge to the unique no-arbitrage cost as n become larger and larger. We now give the details.

Let u and d be as given by Equations (7.1) and (7.2), and let Y be a binomial random variable with parameters n and p, where

$$p = \frac{1 + rt/n - d}{u - d}$$

$$\approx \frac{rt/n + \sigma\sqrt{t/n} - \sigma^2 t/2n}{2\sigma\sqrt{t/n}}$$

$$= \frac{1}{2} + \frac{r\sqrt{t/n}}{2\sigma} - \frac{\sigma\sqrt{t/n}}{4}.$$

It follows from the results of Section 6.2 that the unique no-arbitrage option cost for this n-period model is

$$C = (1 + rt/n)^{-n} E[(S(0)u^Y d^{n-Y} - K)^+]$$

$$= (1 + rt/n)^{-n} E\left[\left(S(0)\left(\frac{u}{d}\right)^Y d^n - K\right)^+\right]$$

$$= (1 + rt/n)^{-n} E[(S(0)e^{2\sigma\sqrt{t/n}\,Y} e^{-\sigma\sqrt{nt}} - K)^+]$$

$$= (1 + rt/n)^{-n} E[(S(0)e^W - K)^+], \tag{7.3}$$

where

$$W = 2\sigma\sqrt{t/n}\,Y - \sigma\sqrt{nt}.$$

Since Y is a binomial random variable with parameters n and p, it follows that, as n becomes larger, Y becomes a normal random variable. Also, since a constant plus a constant multiple of a normal random variable is also normal, it follows that, as n becomes larger, W also becomes a normal random variable. In addition, since $E[Y] = np$,

$$E[W] = 2\sigma\sqrt{t/n}\,E[Y] - \sigma\sqrt{nt}$$

$$= 2\sigma\sqrt{t/n}\,np - \sigma\sqrt{nt}$$

$$= 2\sigma\sqrt{nt}\,(p - 1/2)$$

$$\approx 2\sigma\sqrt{nt}\left(\frac{r\sqrt{t/n}}{2\sigma} - \frac{\sigma\sqrt{t/n}}{4}\right)$$

$$= (r - \sigma^2/2)t. \tag{7.4}$$

Moreover, $\text{Var}(Y) = np(1 - p)$ and $p \approx 1/2$ for large n, so we have that

$$\mathrm{Var}(W) = (2\sigma\sqrt{t/n})^2\,\mathrm{Var}(Y)$$
$$= 4\sigma^2 tp(1-p)$$
$$\approx \sigma^2 t. \tag{7.5}$$

Because all approximations become exact as n grows larger, we see from Equations (7.3)–(7.5) that C, the unique cost of the option that does not result in an arbitrage when the underlying security's price follows a geometric Brownian motion with volatility parameter σ, is

$$C = e^{-rt}E[(S(0)e^W - K)^+], \tag{7.6}$$

where W is a normal random variable with mean $(r - \sigma^2/2)t$ and variance $\sigma^2 t$.

Using standard formulas for normal probabilities, the preceding expression for C can be evaluated to give the following, known as the *Black–Scholes option pricing formula:*

$$C = S(0)\Phi(\omega) - Ke^{-rt}\Phi(\omega - \sigma\sqrt{t}), \tag{7.7}$$

where

$$\omega = \frac{rt + \sigma^2 t/2 - \log(K/S(0))}{\sigma\sqrt{t}}$$

and where $\Phi(x)$ is the standard normal distribution function.

Remarks. (1) Another way of deriving the no-arbitrage option cost C is by finding probabilities on the prices that make all security-buying bets "fair," in the sense that the expected present value of the amount gained is zero for every such bet. Since, as shown in Section 6.2, there is a unique set of probabilities in each of the n-period approximation models, it is reasonable to suppose the same is true for geometric Brownian motion. However, it can be shown that all security-buying bets will be fair if the price of the security follows a geometric Brownian motion with parameters μ and σ, when

$$\mu = r - \sigma^2/2. \tag{7.8}$$

Hence, it follows from the arbitrage theorem that – in order for there not to be an arbitrage when one is also allowed to buy or sell the option – the cost of the option must equal the present value of its expected worth at time t, where the expected worth is computed under the assumption that the parameters of the underlying geometric Brownian motion satisfy Equation (7.8). Consequently, the unique no-arbitrage cost of an option to purchase the stock at time t for the specified strike price K is

$$C = e^{-rt}E[(S(t) - K)^+],$$

where $S(t)/S(0)$ is lognormal with parameters $(r - \sigma^2/2)t$ and $\sigma^2 t$. However, letting

$$W = \log(S(t)/S(0))$$

shows again that

$$C = e^{-rt}E[(S(0)e^W - K)^+],$$

where W is normal with mean $(r - \sigma^2/2)t$ and variance $\sigma^2 t$.

(2) If an investor is neutral toward risk in the sense that she values investments solely by their expected returns, and if she assumes that the underlying security follows a geometric Brownian motion that makes all security-buying and -selling bets fair, then her valuation of the cost of the option would be precisely as given by the Black–Scholes formula. As a result, the Black–Scholes valuation is often called a *risk-neutral valuation*.

(3) Let $C(s, t, K)$ be the no-arbitrage cost of an option having exercise price K and exercise time t when the initial price is s. That is, $C(s, t, K)$ is the C of the Black–Scholes formula having $S(0) = s$. If, at time y $(0 < y < t)$, the price of the underlying security is $S(y) = s_y$, then $C(s_y, t - y, K)$ is the unique no-arbitrage cost of the option at time y. This follows because, at time y, the option will expire after an additional time $t - y$ with the same exercise price K, and for the next $t - y$ units of time the security will follow a geometric Brownian motion with initial value s_y.

(4) It follows from the put–call option parity formula given in Proposition 5.2.2 that the no-arbitrage cost of a European put option with initial price s, strike price K, and exercise time t is given by

$$P(s, t, K) = C(s, t, K) + Ke^{-rt} - s,$$

where $C(s, t, K)$ is the no-arbitrage cost of a call option on the same stock.

(5) The rate of change in the value of the call option as a function of a change in the price of the underlying security is described by the quantity *delta,* denoted as Δ. Formally, if $C(s, t, K)$ is the Black–Scholes cost valuation of the option then Δ is its partial derivative with respect to s; that is,

$$\Delta = \frac{\partial}{\partial s} C(s, t, K).$$

Using Equation (7.7), we can show that

$$\Delta = \Phi(\omega),$$

where ω is as given in that equation. Delta can be used to construct investment portfolios that hedge against risk. For instance, suppose an investor feels that a call option is underpriced and consequently buys the call. To protect himself against a decrease in its price, he can simultaneously sell a certain number of shares of the security. To determine how many shares he should sell, we note that if the price of the security decreases by the small amount h then the worth of the option will decrease by the amount $h\Delta$, implying that the investor would be covered if he had sold Δ shares of the security. Therefore, a reasonable hedge might be to sell Δ shares of the security for each option purchased.

Example 7.1a Suppose that a security is presently selling for a price of 30, the nominal interest rate is 8% (with the unit of time being one year), and the security's volatility is .20. Find the no-arbitrage cost of a call option expiring in three months with a strike price of 34.

Solution. The parameters are

$$t = .25, \quad r = .08, \quad \sigma = .20, \quad K = 34, \quad S(0) = 30,$$

so we have that

$$\omega = \frac{.02 + .005 - \log(34/30)}{(.2)(.5)} \approx -1.0016.$$

Therefore,

$$C = 30\Phi(-1.0016) - 34e^{-.02}\Phi(-1.1016)$$

$$= 30(.15827) - 34(.9802)(.13532)$$

$$\approx .2383.$$

Thus, the appropriate price of the option is 24 cents.   □

7.2   Properties of Black–Scholes Option Cost

The no-arbitrage option cost $C = C(s, t, K, \sigma, r)$ is a function of five variables: the security's initial price s; the exercise time t of the option; the strike price K; the security's volatility parameter σ; and the interest rate r. We will now see what happens to the cost as each of these variables increases. However, as a preliminary to doing so, note first that a normal random variable W with mean $(r - \sigma^2/2)t$ and variance $\sigma^2 t$ can be expressed as

$$W = rt - \sigma^2 t/2 + \sigma\sqrt{t}Z,$$

where Z is a standard normal random variable with mean 0 and variance 1. Hence, using the representation of C as given by Equation (7.6), we see that

$$C(s, t, K, \sigma, r) = e^{-rt}E[(se^W - K)^+]$$
$$= E[(se^{W-rt} - Ke^{-rt})^+], \qquad (7.9)$$

$$C(s, t, K, \sigma, r) = E[(se^{-\sigma^2 t/2 + \sigma\sqrt{t}Z} - Ke^{-rt})^+]. \qquad (7.10)$$

The properties of $C = C(s, t, K, \sigma, r)$ may be described as follows.

C is increasing in s. This means that if the other four variables remain the same, then the cost of the option increases when the initial price increases. This very intuitive result immediately follows from the representation (7.9). For since W does not depend on s, the quantity $(se^{W-rt} - Ke^{-rt})^+$ increases as s increases, and therefore so does its expected value, which by (7.9) is equal to C.

C is decreasing in K. This extremely intuitive result, that the value of the option is larger when the strike price is smaller, also immediately follows from Equation (7.9) since $(se^{W-rt} - Ke^{-rt})^+$ is decreasing in K.

C is increasing in t. Although a mathematical argument can be given, it requires some work. A simpler (and more intuitive) argument is obtained by noting how we can see immediately that the option cost would be increasing in t if the option were an *American* call option (for any additional time to exercise could not hurt, since one could always elect not to use it). But, since the value of a European call option is the same as that of an American call option (Proposition 5.2.1), the result follows.

C is increasing in σ. Because an option holder will greatly benefit from very large prices at the exercise time (while any additional price decrease below the exercise price will not cause any additional loss), this result seems at first sight to be quite intuitive. However, it is more subtle than it appears, since an increase in σ results – not only in an increase in the variance of the logarithm of the final price under the risk-neutral valuation – but also in a *decrease* in the mean (since $E[\log(S(t)/S(0))] = (r - \sigma^2/2)t$). Nevertheless, the result can be proved mathematically.

C is increasing in r. Although this property follows directly from Equation (7.10), since $(se^{-\sigma^2 t/2 + \sigma\sqrt{t}Z} - Ke^{-rt})^+$ is increasing in r, it is probably the least intuitive of the five properties.

7.3   Estimating σ

In order to use the Black–Scholes option cost formula given by Equation (7.7), one must know the value of the parameter σ of the underlying geometric Brownian motion. As this value is initially unknown, historical data must be used to estimate it. So, suppose that we have n consecutive days of data concerning the security. Let P_0 be the closing price of the security immediately before these n days, and let P_k ($k = 1, \ldots, n$) be the closing price at the end of the kth day. If we now set

$$X_k = \log(P_k) - \log(P_{k-1}),$$

then it follows (under the geometric Brownian motion model) that $X_1, \ldots, X_n$ is a sequence of independent normal random variables with a common mean and with common variance $\sigma^2/252$. Here we have taken one year as the unit of time and have used the fact that there are approximately 252 business days in a year (so one day represents approximately $1/252$ of the trading days in a year). We can now apply standard statistical procedures for estimating this variance. Namely, let

$$\bar{X} = \frac{\sum_{k=1}^{n} X_k}{n}$$

be the average of the values of the X_k, and then set

$$V^2 = \frac{\sum_{k=1}^{n}(X_k - \bar{X})^2}{n - 1}.$$

The quantity V^2 is called the *sample variance,* and its value can be taken as the estimate of $\sigma^2/252$. Consequently, $\sqrt{252}V$ is the estimated value of σ.

7.4   Pricing American Put Options

There is no difficulty in determining the risk-neutral prices of European put options; using the put–call option parity formula, it follows that

$$P(s, t, K, \sigma, r) = C(s, t, K, \sigma, r) + Ke^{-rt} - s,$$

where $P(s, t, K, \sigma, r)$ is the risk-neutral price of a European put having strike price K at exercise time t if the price at time 0 is s, the volatility of the stock is σ, the interest rate is r, and $C(s, t, K, \sigma, r)$ is the corresponding risk-neutral price for the call option. However, because early exercise is sometimes beneficial, the risk-neutral pricing of American put options is not so straightforward. We will now present an efficient technique for obtaining very accurate approximations of these prices.

The risk-neutral price of an American put option is the expected present value of owning the option under the assumption that the prices of the underlying security change in accordance with risk-neutral geometric Brownian motion and that the owner utilizes an optimal policy in determining when, if ever, to exercise that option. To approximate this price, we approximate the risk-neutral geometric Brownian motion process by a multiperiod binomial process as follows. Choose a number n and, with t equal to the exercise time of the option, let $t_k = kt/n$ ($k = 0, 1, \ldots, n$). Now suppose that:

(1) the option can be exercised only at one of the times t_k ($k = 0, 1, \ldots, n$), and
(2) if $S(t_k)$ is the price of the security at time t_k, then

$$S(t_{k+1}) = \begin{cases} uS(t_k) & \text{with probability } p, \\ dS(t_k) & \text{with probability } 1 - p, \end{cases}$$

where

$$u = e^{\sigma\sqrt{t/n}}, \qquad d = e^{-\sigma\sqrt{t/n}},$$

$$p = \frac{1 + rt/n - d}{u - d}.$$

The first two possible price movements of this process are indicated in Figure 7.1.

We know from Section 7.1 that the preceding discrete time approximation becomes the risk-neutral geometric Brownian motion process as n

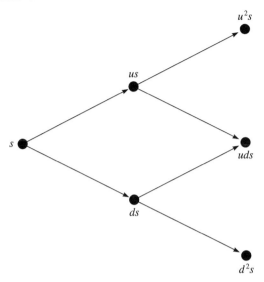

Figure 7.1: Possible Prices of the Discrete Approximation Model

becomes larger and larger; in addition, since the price curve under geometric Brownian motion can be shown to be continuous, it is intuitive (and can be verified) that the expected loss – incurred in allowing the option to be exercised only at one of the times t_k – goes to 0 as n becomes larger. Hence, by choosing n reasonably large, the risk-neutral price of the American option can be accurately approximated by the expected present value return from the option, assuming that both conditions (1) and (2) hold and also that an optimal policy is employed in determining when to exercise the option. We now show how to determine this expected return.

To start, note that if i of the first k price movements were increases and $k - i$ were decreases, then the price at time t_k would be

$$S(t_k) = u^i d^{k-i} s.$$

Since i must be one of the values $0, 1, \ldots, k$, it follows that there are $k + 1$ possible prices of the security at time t_k. Now, let $V_k(i)$ denote the time-t_k expected return from the put, given that the put has not been exercised before time t_k, that the price at time t_k is $S(t_k) = u^i d^{k-i} s$, and that an optimal policy will be followed from time t_k onward.

To determine $V_0(0)$, the expected present value return of owning the put, we work backwards. That is, first we determine $V_n(i)$ for each of its $n + 1$ possible values of i; then we determine $V_{n-1}(i)$ for each of its n possible values of i; then $V_{n-2}(i)$ for each of its $n - 1$ possible values of i; and so on. To accomplish this task, note first that, since the option expires at time t_n, we have

$$V_n(i) = \max(K - u^i d^{n-i} s, \, 0), \tag{7.11}$$

which determines all the values $V_n(i)$, $i = 0, \ldots, n$. Now, let

$$\beta = e^{-rt/n}.$$

Suppose we are at time t_k, the put has not yet been exercised, and the price of the stock is $u^i d^{k-i} s$. If we exercise the option at this point then we will receive $K - u^i d^{k-i} s$. On the other hand, if we do not exercise now then the price at time t_{k+1} will be either $u^{i+1} d^{k-i} s$ (with probability p) or $u^i d^{k-i+1} s$ (with probability $1 - p$). If it is $u^{i+1} d^{k-i} s$ and we employ an optimal policy from that time on, then the time-t_k expected return from the put is $\beta V_{k+1}(i + 1)$; similarly, the expected return if the price decreases is $\beta V_{k+1}(i)$. Hence, since the price will increase with probability p or decrease with probability $1 - p$, it follows that the expected time-t_k return, if we do not exercise but thereafter continue optimally, is

$$p\beta V_{k+1}(i + 1) + (1 - p)\beta V_{k+1}(i).$$

Since $K - u^i d^{k-i} s$ is the return if we exercise and the preceding is the maximal expected return if we do not exercise, it follows that the maximal possible expected return is the larger of these two. That is, for $k = 0, \ldots, n - 1$,

$$V_k(i) = \max\big(K - u^i d^{k-i} s,$$
$$\beta p V_{k+1}(i + 1) + \beta(1 - p)V_{k+1}(i)\big), \quad i = 0, \ldots, k. \tag{7.12}$$

To obtain the approximation, we first use Equation (7.11) to determine the values of $V_n(i)$; we then use (7.12) with $k = n - 1$ to obtain the values $V_{n-1}(i)$; we then use (7.12) with $k = n - 2$ to obtain the values

$V_{n-2}(i)$; and so on until we have the desired value of $V_0(0)$, the approximation of the risk-neutral price of the American put option. Although computationally messy when done by hand, this procedure is easily programmed and can also be done with a spreadsheet.

Remark. The computations can be simplified by noting that $ud = 1$ and also by making use of the following results, which can be shown to hold.

(1) If the put is worthless at time t_k when the price of the security is x, then it is also worthless at time t_k when the price of the security is greater than x. That is,

$$V_k(i) = 0 \implies V_k(j) = 0 \quad \text{if } j > i.$$

(2) If it is optimal to exercise the put option at time t_k when the price is x, then it is also optimal to exercise it at time t_k when the price of the security is less than x. That is,

$$V_k(i) = K - u^i d^{k-i} s \implies V_k(j) = K - u^j d^{k-j} s \quad \text{if } j < i.$$

Example 7.4a   Suppose we want to price an American put option having the following parameters:

$$s = 9, \quad t = .25, \quad K = 10, \quad \sigma = .3, \quad r = .06.$$

To illustrate the procedure, suppose we let $n = 5$ (which is much too small for an accurate approximation). With the preceding parameters, we have that

$$u = e^{.3\sqrt{.05}} = 1.0694,$$

$$d = e^{-.3\sqrt{.05}} = .9351,$$

$$p = .5056,$$

$$1 - p = .4944,$$

$$\beta = e^{-rt/n} = .997.$$

The possible prices of the security at time t_5 are:

$$9d^5 = 6.435,$$

$$9ud^4 = 7.359,$$

$$9u^2d^3 = 8.416,$$

$$9u^3d^2 = 9.625,$$

$$9u^id^{5-i} > 10 \quad (i = 4, 5).$$

Hence,

$$V_5(0) = 3.565,$$

$$V_5(1) = 2.641,$$

$$V_5(2) = 1.584,$$

$$V_5(3) = .375,$$

$$V_5(i) = 0 \quad (i = 4, 5).$$

Since $9u^2d^2 = 9$, Equation (7.12) gives

$$V_4(2) = \max(1, \ \beta p V_5(3) + \beta(1-p)V_5(2)) = 1,$$

which shows that it is optimal to exercise the option at time t_4 when the price is 9. From remark (2) it follows that the option should also be exercised at this time at any lower price, so

$$V_4(1) = 10 - 9ud^3 = 2.130$$

and

$$V_4(0) = 10 - 9d^4 = 3.119.$$

As $9u^3d = 10.293$, Equation (7.12) gives

$$V_4(3) = \beta p V_5(4) + \beta(1-p)V_5(3) = .181.$$

Similarly,

$$V_4(4) = \beta p V_5(5) + \beta(1-p)V_5(4) = 0.$$

Continuing, we obtain:

$$V_3(0) = \max(2.641, \ \beta p V_4(1) + \beta(1-p)V_4(0)) = 2.641,$$

$$V_3(1) = \max(1.584, \ \beta p V_4(2) + \beta(1-p)V_4(1)) = 1.584,$$

$$V_3(2) = \max(.375, \ \beta p V_4(3) + \beta(1-p)V_4(2)) = .584,$$

$$V_3(3) = \beta p V_4(4) + \beta(1-p)V_4(3) = .089;$$

$$V_2(0) = \max(2.130, \ \beta p V_3(1) + \beta(1-p)V_3(0)) = 2.130,$$

$$V_2(1) = \max(1, \ \beta p V_3(2) + \beta(1-p)V_3(1)) = 1.075,$$

$$V_2(2) = \beta p V_3(3) + \beta(1-p)V_3(2) = .333;$$

$$V_1(0) = \max(1.584, \ \beta p V_2(1) + \beta(1-p)V_2(0)) = 1.592,$$

$$V_1(1) = \max(.375, \ \beta p V_2(2) + \beta(1-p)V_2(1)) = .698.$$

As a result,

$$V_0(0) = \max(1, \ \beta p V_1(1) + \beta(1-p)V_1(0)) = 1.137.$$

That is, the risk-neutral price of the put option is approximately 1.137. (The exact answer, to three decimal places, is 1.126, indicating a very respectable approximation given the small value of n that was used.)

7.5 Comments

7.5.1 *When the Option Cost Differs from the Black–Scholes Formula*

Suppose now that we have estimated the value of σ and have inserted that value into the Black–Scholes Equation (7.7) to obtain C. What if the market price of the option is unequal to C? Is there really a strategy that yields us a sure win?

Unfortunately, the answer to the preceding question is "probably not." For one thing, the arbitrage strategy when the actual trading price for the option differs from that given by the Black–Scholes formula requires that one continuously trade (buy or sell) the underlying security. Not only is this physically impossible, but even if discretely approximated it might (in practice) result in large transaction costs that could easily exceed the gain of the arbitrage. A second reason for our answer is that even if we are willing to accept that our estimate of the historical value of σ is very precise, it is possible that its value might change over the option's life. Indeed, perhaps one reason that the market price differs from the formula is because "the market" believes that the stock's volatility over the life of the option will not be the same as it was historically. Indeed, it has been suggested that – rather than using historical data to estimate a security's volatility – a more accurate estimate can often be obtained by finding the value of σ that, along with the other parameters (s, t, K, and r) of the option, makes the Black–Scholes valuation equal to the actual market cost of the option. However, one difficulty with this *implied volatility* is that different options on the same security, having either different exercise times or strike prices or both, will often give rise to different implied volatility estimates of σ. A common occurrence is that implied volatilities derived from in-the-money call options (i.e., ones whose present market price exceeds the strike price) are larger than ones derived from out-of-the-money options (where the present price is less than the strike price). With respect to the Black–Scholes valuation based on estimating σ via historical data, these comments suggest that in-the-money call options tend to be overpriced and that out-of-the-money options are underpriced. One hypothesis that would resolve this seeming anomaly is that the market's prediction of a security's future volatility depends on its recent price movements; if it has recently increased, then the market gives a lower prediction of its near future volatility. For consider an out-of-the-money call option under such an hypothesis. In order for this option to have a positive payoff the security's price must rise, but if that occurs then the market predicts that the security's volatility will become smaller, thus decreasing the value of the option.

A third (even more basic) reason why there is probably no way to guarantee a win is that our assumption that the underlying security follows a geometric Brownian motion is only an approximation to reality, and – even ignoring transaction costs – the existence of an arbitrage strategy relies on this assumption. Indeed, many traders would argue against the geometric Brownian motion assumption that future price changes are independent of past prices, claiming to the contrary that past prices are often an indication of an upward or downward trend in future prices.

7.5.2   *When the Interest Rate Changes*

We have previously shown that the option cost is an increasing function of the interest rate. Does this imply that the cost of an option should increase if the central bank announces an increase in the interest rate (say, on U.S. treasuries) and should decrease if the bank announces a decrease in the interest rate? The answer is yes, *provided* that the security's volatility remains the same. However, one should be careful about making this assumption that a security's volatility will remain unchanged when there is a change in interest rates. An increase in interest rates often has the effect of causing some investors to switch from stocks to either bonds or investments having a fixed return rate, with the reverse resulting when there is a decrease in interest rates; such actions will probably result in a change in the volatility of a security.

7.5.3   *Final Comments*

If you believe that geometric Brownian motion is a reasonable (albeit approximate) model, then the Black–Scholes formula gives a reasonable option price. If this price is significantly above (below) the market price, then a strategy involving buying (selling) options and selling (buying) the underlying security can be devised. Such a strategy, although not yielding a certain win, can often yield a gain that has a positive expected value along with a small variance.

Under the assumption that the security prices over time follow a geometric Brownian motion with parameters μ and σ, one can often devise strategies that have positive expected gains and relatively small risks *even when the cost of the option is as given by the Black–Scholes formula*. For suppose that, based on an estimation using empirical data, you believe that the parameter μ is unequal to the risk-neutral value $r - \sigma^2/2$. If

$$\mu > r - \sigma^2/2$$

then both buying the security and buying the call option will result in positive expected present value gains. Although you cannot avoid all risks (since no arbitrage is possible), a low-risk strategy with a positive expected gain can be effected either by (a) introducing a risk-averse utility function and then finding a strategy that maximizes the expected utility, or (b) finding a strategy that has a reasonably large expected gain along with a reasonably small variance. Such strategies would either buy some security shares and sell some calls, or the reverse. Similarly, if

$$\mu < r - \sigma^2/2$$

then both buying the security and buying the call option have negative expected present value gains, and again we can search for a low-risk, positive expectation strategy that sells one and buys the other. These types of problems are considered in Chapter 8, which also introduces utility functions and their uses.

It is our opinion that the geometric Brownian motion model of the prices of a security over time can often be substantially improved upon, and that – rather than blindly assuming such – one could sometimes do better by using historical data to fit a more general model. If successful, the improved model can give more accurate option prices, resulting in more efficient strategies. The final two chapters of this book deal with these more general models. In Chapter 10 we show that geometric Brownian motion is not consistent with actual data on crude oil prices; an improved model that allows tomorrow's closing price to depend not only on today's closing price but also on yesterday's is presented, and a risk-neutral option price valuation based on this model is indicated. In Chapter 11 we show that a generalization of the geometric Brownian motion model results in an autoregressive model that can be used when modeling a security whose prices have a mean reverting quality.

7.6   Exercises

Unless stated otherwise, the unit of time should be taken as one year.

Exercise 7.1   If the volatility of a stock is .33, what is the standard deviation of (a) $\log(S_d(n)/S_d(n-1))$ and (b) $\log(S_m(n)/S_m(n-1))$? Here, $S_d(n)$ and $S_m(n)$ are (resp.) the prices of the security at the end of day n and month n.

Exercise 7.2   The prices of a certain security follow a geometric Brownian motion with parameters $\mu = .12$ and $\sigma = .24$. If the security's price is presently 40, what is the probability that a call option, having four months to exercise time and with a strike price of $K = 42$, will be exercised? (A security whose price at the time of expiration of a call option is above the strike price is said to finish *in the money*.)

Exercise 7.3   If the interest rate is 8%, what is the risk-neutral valuation of the call option specified in Exercise 7.2?

Exercise 7.4 What is the risk-neutral valuation of a six-month European put option to sell a security for a price of 100 when the current price is 105, the interest rate is 10%, and the volatility of the security is .30?

Exercise 7.5 What should the price of a call option be if the strike price is equal to zero?

Exercise 7.6 What value should the cost of a call option approach as the exercise time becomes larger and larger? Explain your thinking (or do the mathematics).

Exercise 7.7 A European *asset-or-nothing* call pays its holder a fixed amount F if the price at expiration time is larger than K and pays 0 otherwise. Find the risk-neutral valuation of such a call – one that expires in six month's time and has $F = 100$ and $K = 40$ – if the present price of the security is 38, its volatility is .32, and the interest rate is 6%.

Exercise 7.8 To determine the probability that a European call option finishes in the money (see Exercise 7.2), is it enough to specify the five parameters $(K, S(0), r, t,$ and $\sigma)$? Explain your answer. If it is "no," what else is needed?

Exercise 7.9 Continue Figure 7.1 so that it gives the possible price patterns for times $t_0, t_1, t_2, t_3,$ and t_4.

Exercise 7.10 Using the notation of Section 7.4, which of the following statements do you think are true? Explain your reasoning.

(a) $V_k(i)$ is nondecreasing in k for fixed i.
(b) $V_k(i)$ is nonincreasing in k for fixed i.
(c) $V_k(i)$ is nondecreasing in i for fixed k.
(d) $V_k(i)$ is nonincreasing in i for fixed k.

Exercise 7.11 Give the risk-neutral price of a European put option whose parameters are as given in Example 7.4a.

Exercise 7.12 Derive an approximation to the risk-neutral price of an American put option having parameters

$$s = 10, \quad t = .25, \quad K = 10, \quad \sigma = .3, \quad r = .06.$$

Exercise 7.13   Explain how you can use the multiperiod binomial model to approximate the risk-neutral geometric Brownian motion price of an American asset-or-nothing call option.

Exercise 7.14   Derive an approximation to the risk-neutral price of an American asset-or-nothing call option having parameters

$$s = 10, \quad t = .25, \quad K = 11, \quad F = 20, \quad \sigma = .3, \quad r = .06.$$

REFERENCES

The Black–Scholes formula was derived in [1] by solving a stochastic differential equation. The idea of obtaining it by approximating geometric Brownian motion using multiperiod binomial models was developed in [2]. References [3], [4], and [5] are popular textbooks that deal with options, although at a higher mathematical level than the present text.

[1] Black, F., and M. Scholes (1973). "The Pricing of Options and Corporate Liabilities." *Journal of Political Economy* 81: 637–59.

[2] Cox, J., S. A. Ross, and M. Rubinstein (1979). "Option Pricing: A Simplified Approach." *Journal of Financial Economics* 7: 229–64.

[3] Cox, J., and M. Rubinstein (1985). *Options Markets.* Englewood Cliffs, NJ: Prentice-Hall.

[4] Hull, J. (1997). *Options, Futures, and Other Derivatives,* 3rd ed. Englewood Cliffs, NJ: Prentice-Hall.

[5] Luenberger, D. (1998). *Investment Science.* Oxford: Oxford University Press.

8. Valuing by Expected Utility

8.1   Limitations of Arbitrage Pricing

Although arbitrage can be a powerful tool in determining the appropriate cost of an investment, it is more the exception than the rule that it will result in a unique cost. Indeed, as the following example indicates, a unique no-arbitrage option cost will not even result in simple one-period option problems if there are more than two possible next-period security prices.

Example 8.1a   Consider the call option example given in Section 5.1. Again, let the initial price of the security be 100, but now suppose that the price at time 1 can be any of the values 50, 200, and 100. That is, we now allow for the possibility that the price of the stock at time 1 is unchanged from its initial price (see Figure 8.1). As in Section 5.1, suppose that we want to price an option to purchase the stock at time 1 for the fixed price of 150.

For simplicity, let the interest rate r equal zero. The arbitrage theorem states that there will be no guaranteed win if there are nonnegative numbers p_{50}, p_{100}, p_{200} that (a) sum to 1 and (b) are such that the expected gains if one purchases either the stock or the option are zero when p_i is the probability that the stock's price at time 1 is i ($i = 50, 100, 200$). Letting G_s denote the gain at time 1 from buying one share of the *stock*, and letting $S(1)$ be the price of that stock at time 1, we have

$$G_s = \begin{cases} 100 & \text{if } S(1) = 200, \\ 0 & \text{if } S(1) = 100, \\ -50 & \text{if } S(1) = 50. \end{cases}$$

Hence,

$$E[G_s] = 100p_{200} - 50p_{50}.$$

Also, if c is the cost of the option, then the gain from purchasing one *option* is

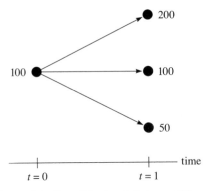

Figure 8.1: Possible Stock Prices at Time 1

$$G_o = \begin{cases} 50 - c & \text{if } S(1) = 200, \\ -c & \text{if } S(1) = 100 \text{ or } S(1) = 50. \end{cases}$$

Therefore,

$$E[G_o] = (50 - c)p_{200} - c(p_{50} + p_{100})$$
$$= 50p_{200} - c.$$

Equating both $E[G_s]$ and $E[G_o]$ to zero shows that the conditions for the absence of arbitrage are that there exist probabilities and a cost c such that

$$p_{200} = \tfrac{1}{2}p_{50} \quad \text{and} \quad c = 50p_{200}.$$

Since the leftmost of the preceding equalities implies that $p_{200} \le 1/3$, it follows that for any value of c satisfying $0 \le c \le 50/3$ we can find probabilities that make both buying the stock and buying the option fair bets. Therefore, no arbitrage is possible for any option cost in the interval $[0, 50/3]$. $\qquad \square$

8.2   Valuing Investments by Expected Utility

Suppose that you must choose one of two possible investments, each of which can result in any of n consequences, denoted $C_1, \ldots, C_n$. Suppose that if the first investment is chosen then consequence i will result with probability p_i $(i = 1, \ldots, n)$, whereas if the second one is chosen then consequence i will result with probability q_i $(i = 1, \ldots, n)$, where $\sum_{i=1}^{n} p_i = \sum_{i=1}^{n} q_i = 1$. The following approach can be used to determine which investment to choose.

We begin by assigning numerical values to the different consequences as follows. First, identify the least and the most desirable consequence, call them c and C respectively; give the consequence c the value 0 and give C the value 1. Now consider any of the other $n - 2$ consequences, say C_i. To value this consequence, imagine that you are given the choice between either receiving C_i or taking part in a random experiment that earns you either consequence C with probability u or consequence c with probability $1 - u$. Clearly your choice will depend on the value of u. If $u = 1$ then the experiment is certain to result in consequence C; since C is the most desirable consequence, you will clearly prefer the experiment to receiving C_i. On the other hand, if $u = 0$ then the experiment will result in the least desirable consequence, namely c, and so in this case you will clearly prefer the consequence C_i to the experiment. Now, as u decreases from 1 down to 0, it seems reasonable that your choice will at some point switch from the experiment to the certain return of C_i, and at that critical switch point you will be indifferent between the two alternatives. Take that indifference probability u as the value of the consequence C_i. In other words, the value of C_i is that probability u such that you are indifferent between either receiving the consequence C_i or taking part in an experiment that returns consequence C with probability u or consequence c with probability $1 - u$. We call this indifference probability the *utility* of the consequence C_i, and we designate it as $u(C_i)$.

In order to determine which investment is superior, we must evaluate each one. Consider the first one, which results in consequence C_i with probability p_i $(i = 1, \ldots, n)$. We can think of the result of this investment as being determined by a two-stage experiment. In the first stage, one of the values $1, \ldots, n$ is chosen according to the probabilities $p_1, \ldots, p_n$; if value i is chosen, you receive consequence C_i. However, since C_i is equivalent to obtaining consequence C with probability $u(C_i)$ or consequence c with probability $1 - u(C_i)$, it follows that the result of the two-stage experiment is equivalent to an experiment in which either consequence C or c is obtained, with C being obtained with probability

$$\sum_{i=1}^{n} p_i u(C_i).$$

Similarly, the result of choosing the second investment is equivalent to taking part in an experiment in which either consequence C or c is obtained, with C being obtained with probability

$$\sum_{i=1}^{n} q_i u(C_i).$$

Since C is preferable to c, it follows that the first investment is preferable to the second if

$$\sum_{i=1}^{n} p_i u(C_i) > \sum_{i=1}^{n} q_i u(C_i).$$

In other words, the value of an investment can be measured by the expected value of the utility of its consequence, and the investment with the largest expected utility is most preferable.

In many investments, the consequences correspond to the investor receiving a certain amount of money. In this case, we let the dollar amount represent the consequence; thus, $u(x)$ is the investor's utility of receiving the amount x. We call $u(x)$ a *utility function*. Thus, if an investor must choose between two investments, of which the first returns an amount X and the second an amount Y, then the investor should choose the first if

$$E[u(X)] > E[u(Y)]$$

and the second if the inequality is reversed, where u is the utility function of that investor. Because the possible monetary returns from an investment often constitute an infinite set, it is convenient to drop the requirement that $u(x)$ be between 0 and 1.

Whereas an investor's utility function is specific to that investor, a general property usually assumed of utility functions is that $u(x)$ is a nondecreasing function of x. In addition, a common (but not universal) feature for most investors is that, if they expect to receive x, then the extra utility gained if they are given an additional amount Δ is nonincreasing in x; that is, for fixed $\Delta > 0$, their utility function satisfies

$$u(x + \Delta) - u(x) \text{ is nonincreasing in } x.$$

A utility function that satisfies this condition is called *concave*. It can be shown that the condition of concavity is equivalent to

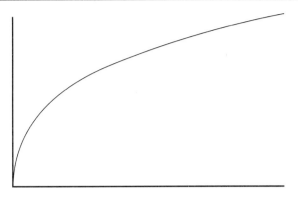

Figure 8.2: A Concave Function

$$u''(x) \leq 0.$$

That is, a function is concave if and only if its second derivative is non-positive. Figure 8.2 gives the curve of a concave function; such a curve always has the property that the line segment connecting any two of its points always lies below the curve.

An investor with a concave utility function is said to be *risk-averse.* This terminology is used because of the following, known as *Jensen's inequality,* which states that if u is a concave function then, for any random variable X,

$$E[u(X)] \leq u(E[X]).$$

Hence, letting X be the return from an investment, it follows from Jensen's inequality that any investor with a concave utility function would prefer the certain return of $E[X]$ to receiving a random return with this mean.

An investor with a linear utility function

$$u(x) = a + bx, \quad b > 0,$$

is said to be *risk-neutral* or *risk-indifferent.* For such a utility function,

$$E[u(X)] = a + bE[X]$$

and so it follows that a risk-neutral investor will value an investment only through its expected return.

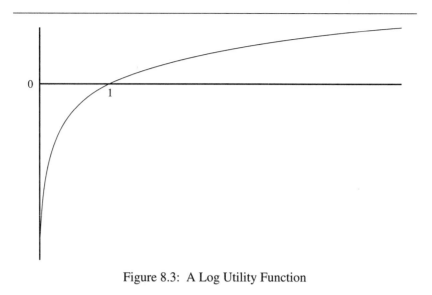

Figure 8.3: A Log Utility Function

A commonly assumed utility function is the *log utility function*

$$u(x) = \log(x);$$

see Figure 8.3. Because $\log(x)$ is a concave function, an investor with a log utility function is risk-averse. This is a particularly important utility function because it can be mathematically proven in a variety of situations that an investor faced with an infinite sequence of investments can maximize long-term rate of return by adopting a log utility function and then maximizing the expected utility in each period. In addition, such a strategy will, for very large values of x, minimize the expected time until the investor's fortune is at least x (see chapter references [1] and [4]). The following example shows how much a log utility investor should invest in a simple favorable gamble.

Example 8.2a   An investor with capital x can invest any amount between 0 and x; if y is invested then y is either won or lost, with respective probabilities p and $1 - p$. If $p > 1/2$, how much should be invested by an investor having a log utility function?

Solution. Suppose the amount αx is invested, where $0 \le \alpha \le 1$. Then the investor's final fortune, call it X, will be either $x + \alpha x$ or $x - \alpha x$

with respective probabilities p and $1 - p$. Hence, the expected utility of this final fortune is

$$p \log((1 + \alpha)x) + (1 - p) \log((1 - \alpha)x)$$
$$= p \log(1 + \alpha) + p \log(x) + (1 - p) \log(1 - \alpha) + (1 - p) \log(x)$$
$$= \log(x) + p \log(1 + \alpha) + (1 - p) \log(1 - \alpha).$$

To find the optimal value of α, we differentiate

$$p \log(1 + \alpha) + (1 - p) \log(1 - \alpha)$$

to obtain

$$\frac{d}{d\alpha}(p \log(1 + \alpha) + (1 - p) \log(1 - \alpha)) = \frac{p}{1 + \alpha} - \frac{1 - p}{1 - \alpha}.$$

Setting this equal to zero yields

$$p - \alpha p = 1 - p + \alpha - \alpha p \quad \text{or} \quad \alpha = 2p - 1.$$

Hence, the investor should always invest $100(2p - 1)$ percent of her present fortune. For instance, if the probability of winning is .6 then the investor should invest 20% of her fortune; if it is .7, she should invest 40%. (When $p \leq 1/2$, it is easy to verify that the optimal amount to invest is 0.) □

Our next example adds a time factor to the previous one.

Example 8.2b Suppose in Example 8.2a that, whereas the investment αx must be immediately paid, the payoff of $2\alpha x$ (if it occurs) does not take place until after one period has elapsed. Suppose further that whatever amount is not invested can be put in a bank to earn interest at a rate of r per period. Now, how much should be invested?

Solution. An investor who invests αx and puts the remaining $(1 - \alpha)x$ in the bank will, after one period, have $(1 + r)(1 - \alpha)x$ in the bank, and the investment will be worth either $2\alpha x$ (with probability p) or 0 (with probability $1 - p$). Hence, the expected value of the utility of his fortune is

$$p \log((1 + r)(1 - \alpha)x + 2\alpha x) + (1 - p) \log((1 + r)(1 - \alpha)x)$$
$$= \log(x) + p \log(1 + r + \alpha - \alpha r)$$
$$+ (1 - p) \log(1 + r) + (1 - p) \log(1 - \alpha).$$

Hence, once again the optimal fraction of one's fortune to invest does not depend on the amount of that fortune. Differentiating the previous equation yields

$$\frac{d}{d\alpha}(\text{expected utility}) = \frac{p(1 - r)}{1 + r + \alpha - \alpha r} - \frac{1 - p}{1 - \alpha}.$$

Setting this equal to zero and solving yields that the optimal value of α is given by

$$\alpha = \frac{p(1 - r) - (1 - p)(1 + r)}{1 - r} = \frac{2p - 1 - r}{1 - r}.$$

For instance, if $p = .6$ and $r = .05$ then, although the expected rate of return on the investment is 20% (whereas the bank pays only 5%), the optimal fraction of money to be invested is

$$\alpha = \frac{.15}{.95} \approx .158.$$

That is, the investor should invest approximately 15.8% of his capital and put the remainder in the bank.   □

Another commonly used utility function is the *exponential utility function*

$$u(x) = 1 - e^{-bx}, \quad b > 0.$$

The exponential is also a risk-averse utility function (see Figure 8.4).

8.3   The Portfolio Selection Problem

Suppose one has the positive amount w to be invested among n different securities. If the amount a is invested in security i ($i = 1, \ldots, n$) then, after one period, that investment returns aX_i, where X_i is a nonnegative random variable. In other words, if we let R_i be the the rate of return from investment i, then

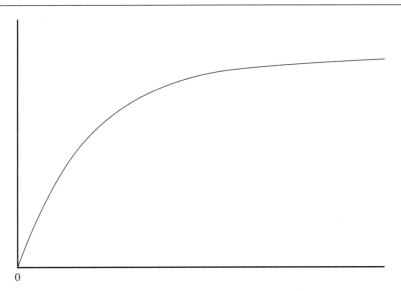

Figure 8.4: An Exponential Utility Function

$$a = \frac{aX_i}{1 + R_i} \quad \text{or} \quad R_i = X_i - 1.$$

If w_i is invested in each security $i = 1, \ldots, n$, then the end-of-period wealth is

$$W = \sum_{i=1}^{n} w_i X_i.$$

The vector $w_1, \ldots, w_n$ is called a *portfolio.* The problem of determining the portfolio that maximizes the expected utility of one's end-of-period wealth can be expressed mathematically as follows:

choose $w_1, \ldots, w_n$ satisfying

$$w_i \geq 0, \quad i = 1, \ldots, n, \quad \sum_{i=1}^{n} w_i = w,$$

to

maximize $E[U(W)]$,

where U is the investor's utility function for the end-of-period wealth.

To make the preceding problem more tractable, we shall make the assumption that the end-of-period wealth W can be thought of as being a normal random variable. Provided that one invests in many securities that are not too highly correlated, this would appear to be, by the central limit theorem, a reasonable approximation. (It would also be exactly true if the X_i, $i = 1, \ldots, n$, have what is known as a multivariate normal distribution.)

Suppose now that the investor has an exponential utility function

$$U(x) = 1 - e^{-bx}, \quad b > 0,$$

and so the utility function is concave. If Z is a normal random variable, then e^Z is lognormal and has expected value

$$E[e^Z] = \exp\{E[Z] + \text{Var}(Z)/2\}.$$

Hence, as $-bW$ is normal with mean $-bE[W]$ and variance $b^2 \, \text{Var}(W)$, it follows that

$$E[U(W)] = 1 - E[e^{-bW}] = 1 - \exp\{-bE[W] + b^2 \, \text{Var}(W)/2\}.$$

Therefore, the investor's expected utility will be maximized by choosing a portfolio that

$$\text{maximizes} \quad E[W] - b \, \text{Var}(W)/2.$$

Observe how this implies that, if two portfolios give rise to random end-of-period wealths W_1 and W_2 such that W_1 has a larger mean and a smaller variance than does W_2, then the first portfolio results in a larger expected utility than does the second. That is,

$$E[W_1] \geq E[W_2] \ \& \ \text{Var}(W_1) \leq \text{Var}(W_2)$$
$$\implies E[U(W_1)] \geq E[U(W_2)]. \tag{8.1}$$

In fact, provided that all end-of-period fortunes are normal random variables, (8.1) remains valid even when the utility function is not exponential, provided that it is a nondecreasing and concave function. Consequently, if one investment portfolio offers a risk-averse investor an expected return that is at least as large as that offered by a second investment portfolio and with a variance that is no greater than that of the second portfolio, then the investor would prefer the first portfolio.

Let us now compute, for a given portfolio, the mean and variance of W. With security i's rate of return $R_i = X_i - 1$, let

$$r_i = E[R_i], \qquad v_i^2 = \mathrm{Var}(R_i).$$

Then, since

$$W = \sum_{i=1}^{n} w_i(1 + R_i) = w + \sum_{i=1}^{n} w_i R_i,$$

we have that

$$E[W] = w + \sum_{i=1}^{n} E[w_i R_i]$$

$$= w + \sum_{i=1}^{n} w_i r_i; \qquad (8.2)$$

$$\mathrm{Var}(W) = \mathrm{Var}\left(\sum_{i=1}^{n} w_i R_i\right)$$

$$= \sum_{i=1}^{n} \mathrm{Var}(w_i R_i)$$

$$+ \sum_{i=1}^{n} \sum_{j \neq i} \mathrm{Cov}(w_i R_i, w_j R_j) \qquad \text{(by Equation (1.11))}$$

$$= \sum_{i=1}^{n} w_i^2 v_i^2 + \sum_{i=1}^{n} \sum_{j \neq i} w_i w_j c(i, j), \qquad (8.3)$$

where

$$c(i, j) = \mathrm{Cov}(R_i, R_j).$$

Example 8.3a Suppose you are thinking about investing your fortune of 100 in two securities whose rates of return have the following expected values and standard deviations:

$$r_1 = .15, \ v_1 = .20; \quad r_2 = .18, \ v_2 = .25.$$

If the correlation between the rates of return is $\rho = -.4$, find the optimal portfolio when employing the utility function

$$U(x) = 1 - e^{-.005x}.$$

Solution. If $w_1 = y$ and $w_2 = 100 - y$, then from Equation (8.2) we obtain

$$E[W] = 100 + .15y + .18(100 - y) = 118 - .03y.$$

Also, since $c(1, 2) = \rho v_1 v_2 = -.02$, Equation (8.3) gives

$$\text{Var}(W) = y^2(.04) + (100 - y)^2(.0625) - 2y(100 - y)(.02)$$
$$= .1425y^2 - 16.5y + 625.$$

We should therefore choose y to maximize

$$118 - .03y - .005(.1425y^2 - 16.5y + 625)/2$$

or, equivalently, to maximize

$$.01125y - .0007125y^2/2.$$

Simple calculus shows that this will be maximized when

$$y = \frac{.01125}{.0007125} = 15.789.$$

That is, the maximal expected utility of the end-of-period wealth is obtained by investing 15.789 in investment 1 and 84.211 in investment 2. Substituting the value $y = 15.789$ into the previous equations gives $E[W] = 117.526$ and $\text{Var}(W) = 400.006$, with the maximal expected utility being

$$1 - \exp\{-.005(117.526 + .005(400.006)/2)\} = .4416.$$

This can be contrasted with the expected utility of .3904 obtained when all 100 is invested in security 1 or the expected utility of .4413 when all 100 is invested in security 2.   □

Example 8.3b   Suppose only two securities are under consideration, both with the same expected rate of return. Then, since every portfolio will yield the same expected return, it follows that the best portfolio for any concave utility function is the one whose end-of-period wealth has minimal variance. If αw is invested in security 1 and $(1 - \alpha)w$ is invested in security 2, then with $c = c(1, 2)$ we have

$$\text{Var}(W) = \alpha^2 w^2 v_1^2 + (1-\alpha)^2 w^2 v_2^2 + 2\alpha(1-\alpha)w^2 c$$
$$= w^2[\alpha^2 v_1^2 + (1-\alpha)^2 v_2^2 + 2c\alpha(1-\alpha)].$$

Thus, the optimal portfolio is obtained by choosing the value of α that minimizes $\alpha^2 v_1^2 + (1-\alpha)^2 v_2^2 + 2c\alpha(1-\alpha)$. Differentiating this quantity and setting the derivative equal to zero yields

$$2\alpha v_1^2 - 2(1-\alpha)v_2^2 + 2c - 4c\alpha = 0.$$

Solving for α gives the optimal fraction to invest in security 1:

$$\alpha = \frac{v_2^2 - c}{v_1^2 + v_2^2 - 2c}.$$

For instance, suppose the standard deviations of the rate of returns are $v_1 = .20$ and $v_2 = .30$, and that the correlation between the two rates of return is $\rho = .30$. Then, as $c = \rho v_1 v_2 = .018$, we obtain that the optimal fraction of one's investment capital to be used to purchase security 1 is

$$\alpha = \frac{.09 - .018}{.04 + .09 - .036} = 72/94 \approx .766.$$

That is, 76.6% of one's capital should be used to purchase security 1 and 23.4% to purchase security 2.

If the rates of returns are independent, then $c = 0$ and the optimal fraction to invest in security 1 is

$$\alpha = \frac{v_2^2}{v_1^2 + v_2^2} = \frac{1/v_1^2}{1/v_1^2 + 1/v_2^2}.$$

In this case, the optimal percentage of capital to invest in a security is determined by a weighted average, where the weight given to a security is inversely proportional to the variance of its rate of return. This result also remains true when there are n securities whose rates of return are uncorrelated and have equal means. Under these conditions, the optimal fraction of one's capital to invest in security i is

$$\frac{1/v_i^2}{\sum_{j=1}^{n} 1/v_j^2}. \qquad \square$$

Determining a portfolio that maximizes the expected utility of one's end-of-period wealth can be computationally quite demanding. Often a reasonable approximation can be obtained when the utility function $U(x)$ satisfies the condition that its second derivative is a nondecreasing function – that is, when

$$U''(x) \text{ is nondecreasing in } x. \tag{8.4}$$

It is easily checked that the utility functions

$$U(x) = x^a, \quad 0 < a < 1,$$
$$U(x) = 1 - e^{-bx}, \quad b > 0,$$
$$U(x) = \log(x)$$

all satisfy the condition of Equation (8.4).

Assuming that $U(x)$ satisfies condition (8.4), we can approximate $U(W)$ by using the first three terms of its Taylor series expansion about the point $\mu = E[W]$. That is, we use the approximation

$$U(W) \approx U(\mu) + U'(\mu)(W - \mu) + U''(\mu)(W - \mu)^2/2.$$

Taking expectations gives that

$$E[U(W)] \approx U(\mu) + U'(\mu)E[W - \mu] + U''(\mu)E[(W - \mu)^2]/2$$
$$= U(\mu) + U''(\mu)v^2/2,$$

where $v^2 = \text{Var}(W)$ and where we have used that

$$E[W - \mu] = E[W] - \mu = \mu - \mu = 0.$$

Therefore, a reasonable approximation to the optimal portfolio is given by the portfolio that maximizes

$$U(E[W]) + U''(E[W]) \text{Var}(W)/2. \tag{8.5}$$

If U is a nondecreasing, concave function that also satisfies condition (8.4), then expression (8.5) will have the desired property of being both increasing in $E[W]$ and decreasing in $\text{Var}(W)$.

Utility functions of the form $U(x) = x^a$ or $U(x) = \log(x)$ have the property that there is a vector

$$\alpha_1^*, \dots, \alpha_n^*, \quad \alpha_i^* \geq 0, \quad \sum_{i=1}^{n} \alpha_i^* = 1,$$

such that the optimal portfolio under a specified one of these utility functions is $w\alpha_1^*, \ldots, w\alpha_n^*$ for *every* initial wealth w. That is, for these utility functions, the optimal proportion of one's wealth w that should be invested in security i does not depend on w. To verify this, note that

$$W = w \sum_{i=1}^{n} \alpha_i X_i$$

for any portfolio $w\alpha_1, \ldots, w\alpha_n$. Hence, if $U(x) = x^a$ then

$$E[U(W)] = E[W^a]$$

$$= E\left[w^a \left(\sum_{i=1}^{n} \alpha_i X_i \right)^a \right]$$

$$= w^a E\left[\left(\sum_{i=1}^{n} \alpha_i X_i \right)^a \right]$$

and so the optimal α_i $(i = 1, \ldots, n)$ do not depend on w. (The argument for $U(x) = \log(x)$ is left as an exercise.)

An important feature of the approximation criterion (8.5) is that, when $U(x) = x^a$ $(0 < a < 1)$, the portfolio that maximizes (8.5) also has the property that the percentage of wealth it invests in each security does not depend on w. This follows since equations (8.2) and (8.3) show that, for the portfolio $w_i = \alpha_i w$ $(i = 1, \ldots, n)$,

$$E[W] = wA, \qquad \text{Var}(W) = w^2 B,$$

where

$$A = 1 + \sum_{i=1}^{n} \alpha_i r_i,$$

$$B = \sum_{i=1}^{n} \alpha_i^2 v_i^2 + \sum_{i=1}^{n} \sum_{j \neq i} \alpha_i \alpha_j c(i, j).$$

Thus, since

$$U''(x) = a(a - 1)x^{a-2},$$

we see that

$$U(E[W]) + U''(E[W]) \, \text{Var}(W)/2$$

$$= w^a A^a + a(a - 1)w^{a-2} A^{a-2} w^2 B/2$$

$$= w^a [A^a + a(a - 1)A^{a-2}B/2].$$

Therefore, the investment percentages that maximize (8.5) do not depend on w.

Example 8.3c   Let us reconsider Example 8.3a, this time using the utility function

$$U(x) = \sqrt{x}.$$

Then, with $\alpha_1 = \alpha$ and $\alpha_2 = 1 - \alpha$ we have

$$A = 1 + .15\alpha + .18(1 - \alpha),$$

$$B = .04\alpha^2 + .0625(1 - \alpha)^2 - 2(.02)\alpha(1 - \alpha),$$

and we must choose the value of α that maximizes

$$f(\alpha) = A^{1/2} - A^{-3/2}B/8.$$

The solution can be obtained by setting the derivative equal to zero and then solving this equation numerically.   □

Suppose now that we can invest a positive or negative amount in any investment and, in addition, that all investments are financed by borrowing money at a fixed rate of r per period. If w_i is invested in investment i ($i = 1, \ldots, n$), then the return from this portfolio after one period is

$$R(\mathbf{w}) = \sum_{i=1}^{n} w_i(1 + R_i) - (1 + r)\sum_{i=1}^{n} w_i = \sum_{i=1}^{n} w_i(R_i - r).$$

(If $s = \sum_i w_i$, then s is borrowed from the bank if $s > 0$ and $-s$ is deposited in the bank if $s < 0$.) Let

$$r(\mathbf{w}) = E[R(\mathbf{w})], \qquad V(\mathbf{w}) = \text{Var}(R(\mathbf{w}))$$

and note that

$$r(a\mathbf{w}) = ar(\mathbf{w}), \qquad V(a\mathbf{w}) = a^2 V(\mathbf{w}),$$

where $a\mathbf{w} = (aw_1, \ldots, aw_n)$. Now, let $\mathbf{w}^*$ be such that $r(\mathbf{w}^*) = 1$ and

$$V(\mathbf{w}^*) = \min_{\mathbf{w}:r(\mathbf{w})=1} V(\mathbf{w}).$$

That is, among all portfolios $\mathbf{w}$ whose expected return is 1, the variance of the portfolio's return is minimized under $\mathbf{w}^*$.

We now show that for any $b > 0$, among all portfolios whose expected return is b, the variance of the portfolio's return is minimized under $b\mathbf{w}^*$. To verify this, suppose that $r(\mathbf{y}) = b$. But then

$$r\left(\frac{1}{b}\mathbf{y}\right) = \frac{1}{b}r(\mathbf{y}) = 1,$$

which implies (by the definition of $\mathbf{w}^*$) that

$$V(b\mathbf{w}^*) = b^2 V(\mathbf{w}^*) \leq b^2 V\left(\frac{1}{b}\mathbf{y}\right) = V(\mathbf{y}),$$

which completes the verification. Hence, portfolios that minimize the variance of the return are constant multiples of a particular portfolio. This is called the portfolio *separation theorem* because, when analyzing the portfolio decision problem from a mean variance viewpoint, the theorem enables us to separate the portfolio decision problem into a determination of the relative amounts to invest in each investment and the choice of the scalar multiple.

8.3.1   *Estimating Covariances*

In order to create good portfolios, we must first use historical data to estimate the values of $r_i = E[R_i]$, $v_i^2 = \text{Var}(R_i)$, and $c(i, j) = \text{Cov}(R_i, R_j)$ for all i and j. The means r_i and variances v_i^2 can be estimated, as was shown in Section 7.3, by using the sample mean and sample variance of historical rates of return for security i. To estimate the covariance $c(i, j)$ for a fixed pair i and j, suppose we have historical data that covers m periods and let $r_{i,k}$ and $r_{j,k}$ denote (respectively) the rates of return of security i and of security j for period k, $k = 1, \ldots, m$. Then, the usual estimator of

$$\text{Cov}(R_i, R_j) = E[(R_i - r_i)(R_j - r_j)]$$

is

$$\frac{\sum_{k=1}^{m}(r_{i,k} - \bar{r}_i)(r_{j,k} - \bar{r}_j)}{m - 1},$$

where $\bar{r}_i$ and $\bar{r}_j$ are the sample means

$$\bar{r}_i = \frac{\sum_{k=1}^{m} r_{i,k}}{m}, \qquad \bar{r}_j = \frac{\sum_{k=1}^{m} r_{j,k}}{m}.$$

8.4   The Capital Assets Pricing Model

The Capital Assets Pricing Model (CAPM) attempts to relate R_i, the one-period rate of return of a specified security i, to R_m, the one-period rate of return of the entire market (as measured, say, by the Standard and Poor's index of 500 stocks). If r_f is the risk-free interest rate (usually taken to be the current rate of a U.S. Treasury bill) then the model assumes that, for some constant β_i,

$$R_i = r_f + \beta_i(R_m - r_f) + e_i,$$

where e_i is a normal random variable with mean 0 that is assumed to be independent of R_m. Letting the expected values of R_i and R_m be r_i and r_m (resp.), the CAPM model (which treats r_f as a constant) implies that

$$r_i = r_f + \beta_i(r_m - r_f)$$

or, equivalently, that

$$r_i - r_f = \beta_i(r_m - r_f).$$

That is, the difference between the expected rate of return of the security and the risk-free interest rate is assumed to equal β_i times the difference between the expected rate of return of the market and the risk-free interest rate. Thus, for instance, if $\beta_i = 1$ (resp. $\frac{1}{2}$ or 2) then the expected amount by which the rate of return of security i exceeds r_f is the same as (resp. one-half or twice) the expected amount by which the overall market's rate of return exceeds r_f. The quantity β_i is known as the *beta* of security i.

Using the linearity property of covariances – along with the result that the covariance of a random variable and a constant is 0 – we obtain from the CAPM that

$$\text{Cov}(R_i, R_m) = \beta_i \, \text{Cov}(R_m, R_m) + \text{Cov}(e_i, R_m)$$
$$= \beta_i \, \text{Var}(R_m) \quad \text{(since } e_i \text{ and } R_m \text{ are independent).}$$

Therefore, letting $v_m^2 = \text{Var}(R_m)$, we see that

$$\beta_i = \frac{\text{Cov}(R_i, R_m)}{v_m^2}.$$

Example 8.4a Suppose that the current risk-free interest rate is 6% and that the expected value and standard deviation of the market rate of return are .10 and .20, respectively. If the covariance of the rate of return of a given stock and the market's rate of return is .05, what is the expected rate of return of that stock?

Solution. Since

$$\beta = \frac{.05}{(.20)^2} = 1.25,$$

it follows (assuming the validity of the CAPM) that

$$r_i = .06 + 1.25(.10 - .06) = .11.$$

That is, the stock's expected rate of return is 11%. □

If we let $v_i^2 = \text{Var}(R_i)$ then under the CAPM it follows, using the assumed independence of R_m and e_i, that

$$v_i^2 = \beta_i^2 v_m^2 + \text{Var}(e_i).$$

If we think of the variance of a security's rate of return as constituting the risk of that security, then the foregoing equation states that the risk of a security is the sum of two terms: the first term, $\beta_i^2 v_m^2$, is called the *systematic risk* and is due to the combination of the security's beta and the inherent risk in the market; the second term, $\text{Var}(e_i)$, is called the *specific risk* and is due to the specific stock being considered.

8.5 Mean Variance Analysis of Risk-Neutral–Priced Call Options

Suppose that the prices of a certain security follow a geometric Brownian motion with parameters μ and σ. Let r be the interest rate, and suppose that

$$\mu \neq r - \sigma^2/2.$$

Furthermore, suppose that a call option to purchase the stock at time t for the price K is selling at the price C specified by the Black–Scholes formula. Then, although there is no sure win, one can still make investments whose present value gain has a positive expectation and a small variance.

To begin, suppose that at time 0 we purchase x_1 units of the security and x_2 units of the option for a total price of $sx_1 + Cx_2$. If we intend to close out the investment at time t, then its present value gain is

$$P = e^{-rt}\left(x_1 S(t) + x_2(S(t) - K)^+\right) - x_1 s - x_2 C.$$

To compute $E[P]$ and $\text{Var}(P)$, let

$$d = \frac{\mu t - \log(K/s)}{\sigma \sqrt{t}}$$

and let

$$F(b) = \exp\{b\mu t + b^2 \sigma^2 t/2\} \Phi(b\sigma \sqrt{t} + d),$$

where Φ is the standard normal distribution function. Now, we let

$$A = sF(1) - KF(0),$$
$$B = s^2 F(2) - KsF(1),$$
$$D = s^2 F(2) - 2KsF(1) + K^2 F(0),$$
$$E = s \exp\{\mu t + t\sigma^2/2\},$$
$$G = s^2 \exp\{2\mu t + 2t\sigma^2\}.$$

It can be shown that

$$A = E[(S(t) - K)^+],$$
$$B = E[S(t)(S(t) - K)^+],$$
$$D = E[((S(t) - K)^+)^2],$$
$$E = E[S(t)],$$
$$G = E[S^2(t)].$$

The preceding yields that

$$E[P] = e^{-rt}(x_1 E + x_2 A) - x_1 s - x_2 C \tag{8.6}$$

and

$$\text{Var}(P) = e^{-2rt} \text{Var}\left(x_1 S(t) + x_2 (S(t) - K)^+\right)$$
$$= e^{-2rt}\left(E\left[\left(x_1 S(t) + x_2 (S(t) - K)^+\right)^2\right]\right.$$
$$- E^2[x_1 S(t) + x_2 (S(t) - K)^+]\right)$$
$$= e^{-2rt}(x_1^2 G + x_2^2 D + 2x_1 x_2 B - (x_1 E + x_2 A)^2). \tag{8.7}$$

One can then experiment with different values of x_1 and x_2, one of which should be positive and the other negative; Equations (8.6) and (8.7) can be used to determine the resulting means and variances. If one can find values of the x_i that give a positive expected gain with an acceptably small variance, then an investment having a relatively small risk can be made.

Example 8.5a   Consider a call option to purchase a security in five months for a price of 60 when the current price of the security is $s = 62$, the volatility of the security is .20 per year, and the interest rate is 10%. In addition, suppose that the call is selling for its Black–Scholes price valuation of $C = 5.80$. If you think that the drift parameter of the geometric Brownian motion that describes the security's price over time is $\mu = .10$ then, as
$$.10 > .10 - (.2)^2/2 = .08,$$

it follows (based on your evaluation) that both the security and the call option have positive expected present value gains. With the notation defined in this section, it turns out that

$$d = .5767, \quad A = 6.4502, \quad B = 476.1476,$$
$$D = 89.0516, \quad E = 65.1788, \quad G = 4,319.387.$$

Consequently, from Equations (8.6) and (8.7) we obtain that

$$E[P] = .5188x_1 + .3870x_2,$$
$$\text{Var}(P) = 65.4253x_1^2 + 43.6529x_2^2 + 102.5505x_1x_2.$$

Therefore, if you buy one share of the stock, then the present value of your gain has expected value $.52 and standard deviation $8.09; on the other hand, if you buy one call option, then your expected present value gain is $.39 with a standard deviation of $6.61. Because buying .746

shares of the security will result in an expected present value gain of $.39 with a standard deviation of $6.04, buying the security is a better investment (in mean-value sense) than is buying the option. However, we can obtain an expected present value gain equal to 1 by letting

$$x_1 = \frac{1 - .3870x_2}{.5188}.$$

If we then choose x_2 to minimize $\text{Var}(P)$, the solution is

$$x_1 = 2.93, \qquad x_2 = -1.34,$$

which results in

$$\sqrt{\text{Var}(P)} = 15.38.$$

(As a comparison, buying $1/.52$ shares of the security results in an expected present value gain of 1 but with a standard deviation of $8.09/.52 = 15.56$.) Thus, the optimal policy (in the sense of minimizing the variance for a specified mean return) is to sell 1.34 options for every 2.93 shares purchased (or, equivalently, to sell $1.34/2.93 \approx .46$ options for every security share purchased).   □

8.6   Rates of Return: Single-Period and Geometric Brownian Motion

Let $S_i(t)$ be the price of security i at time t ($t \geq 0$), and assume that these prices follow a geometric Brownian motion with drift parameter μ_i and volatility parameter σ_i. If R_i is the one-period rate of return for security i, then

$$\frac{S_i(1)}{1 + R_i} = S_i(0)$$

or, equivalently,

$$R_i = \frac{S_i(1)}{S_i(0)} - 1.$$

Since $S_i(1)/S_i(0)$ has the same probability distribution as e^X when X is a normal random variable with mean μ_i and variance σ_i^2, it follows that

$$r_i = E[R_i] = E\left[\frac{S_i(1)}{S_i(0)}\right] - 1$$

$$= E[e^X] - 1$$

$$= \exp\{\mu_i + \sigma_i^2/2\} - 1.$$

Also,

$$v_i^2 = \mathrm{Var}(R_i) = \mathrm{Var}\left(\frac{S_i(1)}{S_i(0)}\right)$$
$$= \mathrm{Var}(e^X)$$
$$= E[e^{2X}] - (E[e^X])^2$$
$$= \exp\{2\mu_i + 2\sigma_i^2\} - (\exp\{\mu_i + \sigma_i^2/2\})^2$$
$$= \exp\{2\mu_i + 2\sigma_i^2\} - \exp\{2\mu_i + \sigma_i^2\},$$

where the next-to-last equality used the fact that $2X$ is normal with mean $2\mu_i$ and variance $4\sigma_i^2$ to determine $E[e^{2X}]$.

Thus, the expected one-period rate of return is $\exp\{\mu_i + \sigma_i^2/2\} - 1$; note that this is *not* the expected value of the average spot rate of return by time 1. For if we let $\bar{R}_i(t)$ be the average spot rate of return by time t (i.e., the yield curve), then

$$\frac{S_i(t)}{S_i(0)} = e^{t\bar{R}_i(t)},$$

implying that

$$\bar{R}_i(t) = \frac{1}{t}\log\left(\frac{S_i(t)}{S_i(0)}\right).$$

Since $\log(S_i(t)/S_i(0))$ is a normal random variable with mean $\mu_i t$ and variance $t\sigma_i^2$, it follows that $\bar{R}_i(t)$ is a normal random variable with

$$E[\bar{R}_i(t)] = \mu_i, \qquad \mathrm{Var}(\bar{R}_i(t)) = \sigma_i^2/t.$$

Thus, the expected value and variance of the one-period yield function for geometric Brownian motion are its parameters μ_i and σ_i^2.

8.7   Exercises

Exercise 8.1   Suppose, in Example 8.1a, that possible prices of the security at time 1 are 50, 175, and 200. Find the range of no-arbitrage option costs. What conclusion can you draw?

Exercise 8.2   The degree of risk aversion indicated by the utility function $u(x)$ is defined as

$$a(x) = -\frac{u''(x)}{u'(x)},$$

where u' and u'' are the first two derivatives of U. The quantity $a(x)$ is called the *Arrow–Pratt absolute risk-aversion coefficient*. Calculate this coefficient when

(a) $u(x) = \log(x)$;
(b) $u(x) = 1 - e^{-x}$.

Exercise 8.3   In Example 8.2a, show that if $p \leq 1/2$ then the optimal amount to invest is 0.

Exercise 8.4   In Example 8.2b, show that if $p \leq 1/2$ then the optimal amount to invest is 0.

Exercise 8.5   Suppose in Example 8.3a that $\rho = 0$. What is the optimal portfolio?

Exercise 8.6   Suppose in Example 8.3a that $r_1 = .16$. Determine the maximal expected utility and compare it with (a) the expected utility obtained when everything is invested in security 1 and (b) the expected utility obtained when everything is invested in security 2.

Exercise 8.7   Show that the percentage of one's wealth that should be invested in each security when attempting to maximize $E[\log(W)]$ does not depend on the amount of initial wealth.

Exercise 8.8   Verify that $U''(x)$ is nondecreasing in x when $x > 0$ and when

(a) $U(x) = x^a, \, 0 < a < 1$;
(b) $U(x) = 1 - e^{-bx}, \, b > 0$;
(c) $U(x) = \log(x)$.

Exercise 8.9   Does the percentage of one's wealth that is to be invested in each security when attempting to maximize the approximation (8.5) depend on initial wealth when $U(x) = \log(x)$?

Exercise 8.10   Use the approximation to $E[U(W)]$ given by (8.5) to determine the optimal amounts to invest in each security in Example 8.3a when using the utility function $U(x) = 1 - e^{-.005x}$. Compare your results with those obtained in that example.

Exercise 8.11 Suppose we want to choose a portfolio with the objective of maximizing the probability that our end-of-period wealth is at least g, where $g > w$. Assuming that W is normal, the optimal portfolio will be the one that maximizes what function of $E[W]$ and $\text{Var}(W)$?

Exercise 8.12 Find the optimal portfolio in Example 8.3a if your objective is to maximize the probability that your end-of-period wealth is at least: (a) 110; (b) 115; (c) 120; (d) 125.

Exercise 8.13 Find the solution of Example 8.3c.

Exercise 8.14 If the beta of a stock is .80, what is the expected rate of return of that stock if the expected value of the market's rate of return is .07 and the risk-free interest rate is 5%? What if the risk-free interest rate is 10%? Assume the CAPM.

Exercise 8.15 If β_i is the beta of stock i for $i = 1, \ldots, k$, what would be the beta of a portfolio in which α_i is the fraction of one's capital that is used to purchase stock i $(i = 1, \ldots, k)$?

Exercise 8.16 A single-factor model supposes that R_i, the one-period rate of return of a specified security, can be expressed as

$$R_i = a_i + b_i F + e_i,$$

where F is a random variable (called the "factor"), e_i is a normal random variable with mean 0 that is independent of F, and a_i and b_i are constants that depend on the security. Show that the CAPM is a single-factor model, and identify a_i, b_i, and F.

Exercise 8.17 In Example 8.5a, find the expected value and the standard deviation of an investment that purchases

(a) 3 shares of the security and -2 options;
(b) 3 options and -2 shares of the security.

REFERENCES

References [2], [3], and [5] deal with utility theory.

[1] Breiman, L. (1960). "Investment Policies for Expanding Businesses Optimal in a Long Run Sense." *Naval Research Logistics Quarterly* 7: 647–51.

[2] Ingersoll, J. E. (1987). *Theory of Financial Decision Making.* Lanham, MD: Rowman & Littlefield.

[3] Pratt, J. (1964). "Risk Aversion in the Small and in the Large," *Econometrica* 32: 122–30.

[4] Thorp, E. O. (1975). "Portfolio Choice and the Kelly Criterion." In W. T. Ziemba and R. G. Vickson (Eds.), *Stochastic Optimization Models in Finance.* New York: Academic Press.

[5] von Neumann, J., and O. Morgenstern (1944). *Theory of Games and Economic Behavior.* Princeton, NJ: Princeton University Press.

9. Exotic Options

9.1 Introduction

The options we have so far considered are sometimes called "vanilla" options to distinguish them from the more exotic options, whose prevalence has increased in recent years. Generally speaking, the value of these options at the exercise time depends not only on the security's price at that time but also on the price path leading to it. In this chapter we introduce three of these exotic-type options – barrier options, Asian options, and lookback options – and show how to use Monte Carlo simulation methods efficently to determine their geometric Brownian motion risk-neutral valuations. In the final section of this chapter we present an explicit formula for the risk-neutral valuation of a "power" call option, whose payoff when exercised is the amount by which a specified power of the security's price at that time exceeds the exercise price.

9.2 Barrier Options

To define a European barrier call option with strike price K and exercise time t, a barrier value v is specified; depending on the type of barrier option, the option either becomes alive or is killed when this barrier is crossed. A *down-and-in* barrier option becomes alive only if the security's price goes below v before time t, whereas a *down-and-out* barrier option is killed if the security's price goes below v before time t. In both cases, v is a specified value that is less than the initial price s of the security. In addition, in most applications, the barrier is considered to be breached only if an end-of-day price is lower than v; that is, a price below v that occurs in the middle of a trading day is not considered to breach the barrier. Now, if one owns both a down-and-in and a down-and-out call option, both with the same values of K and t, then exactly one option will be in play at time t (the down-and-in option if the barrier is breached and the down-and-out otherwise); hence, owning both is equivalent to owning a vanilla option with exercise time t and exercise price K. As a result, if $D_i(s, t, K)$ and $D_o(s, t, K)$ represent, respectively, the risk-neutral present values of owning the down-and-in and the down-and-out call options, then

$$D_i(s, t, K) + D_o(s, t, K) = C(s, t, K),$$

where $C(s, t, K)$ is the Black–Scholes valuation of the call option given by Equation (7.7). As a result, determining either one of the values $D_i(s, t, K)$ or $D_o(s, t, K)$ automatically yields the other.

There are also *up-and-in* and *up-and-out* barrier call options. The up-and-in option becomes alive only if the security's price exceeds a barrier value v, whereas the up-and-out is killed when that event occurs. For these options, the barrier value v is greater than the exercise price K. Since owning both these options (with the same t and K) is equivalent to owning a vanilla option, we have

$$U_i(s, t, K) + U_o(s, t, K) = C(s, t, K),$$

where U_i and U_o are the geometric Brownian motion risk-neutral valuations of (resp.) the up-and-in and the up-and-out call options, and C is again the Black–Scholes valuation.

9.3   Asian and Lookback Options

Asian options are options whose value at the time t of exercise is dependent on the *average price* of the security over at least part of the time between 0 (when the option was purchased) and the time of exercise. As these averages are usually in terms of the end-of-day prices, let N denote the number of trading days in a year (usually taken equal to 252), and let

$$S_d(i) = S(i/N)$$

denote the security's price at the end of day i. The most common Asian-type call option is one in which the exercise time is the end of n trading days, the strike price is K, and the payoff at the exercise time is

$$\left(\sum_{i=1}^{n} \frac{S_d(i)}{n} - K \right)^+.$$

Another Asian option variation is to let the average price be the strike price; the final value of this call option is thus

$$\left(S_d(n) - \sum_{i=1}^{n} \frac{S_d(i)}{n} \right)^+$$

when the exercise time is at the end of trading day n.

Another type of exotic option is the *lookback option,* whose strike price is the minimum end-of-day price up to the option's exercise time. That is, if the exercise time is at the end of n trading days, then the payoff at exercise time is

$$S_d(n) - \min_{i=1,\ldots,n} S_d(i).$$

Because their final payoffs depend on the end-of-day price path followed, there are no known exact formulas for the risk-neutral valuations of barrier, Asian, or lookback options. However, fast and accurate approximations are obtainable from efficient Monte Carlo simulation methods.

9.4   Monte Carlo Simulation

Suppose we want to estimate θ, the expected value of some random variable Y:

$$\theta = E[Y].$$

Suppose, in addition, that we are able to genererate the values of independent random variables having the same probability distribution as does Y. Each time we generate a new value, we say that a simulation "run" is completed. Suppose we perform k simulation runs and so generate the values of (say) $Y_1, Y_2, \ldots, Y_k$. If we let

$$\bar{Y} = \frac{1}{k} \sum_{i=1}^{k} Y_i$$

be their arithmetic average, then $\bar{Y}$ can be used as an estimator of θ. Its expected value and variance are as follows. For the expected value we have

$$E[\bar{Y}] = \frac{1}{k} \sum_{i=1}^{k} E[Y_i] = \theta.$$

Also, letting

$$v^2 = \text{Var}(Y),$$

we have that

$$\text{Var}(\bar{Y}) = \text{Var}\left(\frac{1}{k} \sum_{i=1}^{k} Y_i\right)$$

$$= \frac{1}{k^2} \text{Var}\left(\sum_{i=1}^{k} Y_i\right)$$

$$= \frac{1}{k^2} \sum_{i=1}^{k} \text{Var}(Y_i) \qquad \text{(by independence)}$$

$$= v^2/k.$$

Also, it follows from the central limit theorem that, for large k, $\bar{X}$ will have an approximately normal distribution. Hence, as a normal random variable tends not to be too many standard deviations (equal to the square root of its variance) away from its mean, it follows that if $v/\sqrt{k}$ is small then $\bar{X}$ will tend to be near θ. (For instance, since more than 95% of the time a normal random variable is within two standard deviations of its mean, we can be 95% certain that the generated value of $\bar{X}$ will be within $2v/\sqrt{k}$ of θ.) Hence, when k is large, $\bar{X}$ will tend to be a good estimator of θ. (To know exactly how good, we would use the generated sample variance to estimate v^2.) This approach to estimating an expected value is known as *Monte Carlo simulation*.

9.5   Pricing Exotic Options by Simulation

Suppose that the nominal interest rate is r and that the price of a security follows the risk-neutral geometric Brownian motion; that is, it follows a geometric Brownian motion with variance parameter σ^2 and drift parameter μ, where

$$\mu = r - \sigma^2/2.$$

Let $S_d(i)$ denote the price of the security at the end of day i, and let

$$X(i) = \log\left(\frac{S_d(i)}{S_d(i-1)}\right).$$

Successive daily price ratio changes are independent under geometric Brownian motion, so it follows that $X(1), \ldots, X(n)$ are independent normal random variables, each having mean μ/N and variance σ^2/N (as before, N denotes the number of trading days in a year). Therefore, by generating the values of n independent normal random variables having this mean and variance, we can construct a sequence of n end-of-day prices that have the same probabilities as ones that evolved from the risk-neutral geometric Brownian motion model. (Most computer languages and almost all spreadsheets have built-in utilities for generating the values of standard normal random variables; multiplying these by $\sigma/\sqrt{N}$ and then adding μ/N gives the desired normal random variables.)

Suppose we want to find the risk-neutral valuation of a down-and-in barrier option whose strike price is K, barrier value is v, initial value is $S(0) = s$, and exercise time is at the end of trading day n. We begin by generating n independent normal random variables with mean μ/N and variance σ^2/N. Set them equal to $X(1), \ldots, X(n)$, and then determine the sequence of end-of-day prices from the equations

$$S_d(0) = s,$$

$$S_d(1) = S_d(0)e^{X(1)},$$

$$S_d(2) = S_d(1)e^{X(2)};$$

$$\vdots$$

$$S_d(i) = S_d(i-1)e^{X(i)};$$

$$\vdots$$

$$S_d(n) = S_d(n-1)e^{X(n)}.$$

In terms of these prices, let I equal 1 if an end-of-day price is ever below the barrier v, and let it equal 0 otherwise; that is,

$$I = \begin{cases} 1 & \text{if } S_d(i) < v \text{ for some } i = 1, \ldots, n, \\ 0 & \text{if } S_d(i) \geq v \text{ for all } i = 1, \ldots, n. \end{cases}$$

Then, since the down-and-in call option will be alive only if $I = 1$, it follows that the time-0 value of its payoff at expiration time n is

$$\text{payoff of the down-and-in call option} = e^{-rn/N}I(S_d(n) - K)^+.$$

Call this payoff Y_1. Repeating this procedure an additional $k-1$ times yields $Y_1, \ldots, Y_k$, a set of k payoff realizations. We can then use their average as an estimate of the risk-neutral geometric Brownian motion valuation of the barrier option.

Risk-neutral valuations of Asian and lookback call options are similarly obtained. As in the preceding, we first generate the values of $X(1), \ldots, X(n)$ and use them to compute $S_d(1), \ldots, S_d(n)$. For an Asian option, we then let

$$Y = e^{-rn/N} \left(\sum_{i=1}^{n} \frac{S_d(i)}{n} - K \right)^{+}$$

if the strike price is fixed at K and the payoff is based on the average end-of-day price, or we let

$$Y = e^{-rn/N} \left(S_d(n) - \sum_{i=1}^{n} \frac{S_d(i)}{n} \right)^{+}$$

if the average end-of-day price is the strike price. In the case of a lookback option, we would let

$$Y = e^{-rn/N} \left(S_d(n) - \min_{i} S_d(i) \right).$$

Repeating this procedure an additional $k-1$ times and then taking the average of the k values of Y yields the Monte Carlo estimate of the risk-neutral valuation.

9.6   More Efficient Simulation Estimators

In this section we show how the simulation of valuations of Asian and lookback options can be made more efficient by the use of control and antithetic variables, and how the valuation simulations of barrier options can be improved by a combination of the variance reduction simulation techniques of conditional expectation and importance sampling.

9.6.1   *Control and Antithetic Variables in the Simulation of Asian and Lookback Option Valuations*

Consider the general setup where one plans to use simulation to estimate

$$\theta = E[Y].$$

Suppose that, in the course of generating the value of the random variable Y, we also learn the value of a random variable V whose mean value is known to be $\mu_V = E[V]$. Then, rather than using the value of Y as the estimator, we can use one of the form

$$Y + c(V - \mu_V),$$

where c is a constant to be specified. That this quantity also estimates θ follows by noting that

$$E[Y + c(V - \mu_V)] = E[Y] + cE[V - \mu_V] = \theta + c(\mu_V - \mu_V) = \theta.$$

The best estimator of this type is obtained by choosing c to be the value that makes $\text{Var}(Y + c(V - \mu_V))$ as small as possible. Now,

$$
\begin{aligned}
\text{Var}(Y + c(V - \mu_V)) &= \text{Var}(Y + cV) \\
&= \text{Var}(Y) + \text{Var}(cV) + 2\,\text{Cov}(Y, cV) \\
&= \text{Var}(Y) + c^2\,\text{Var}(V) + 2c\,\text{Cov}(Y, V). \quad (9.1)
\end{aligned}
$$

If we differentiate Equation (9.1) with respect to c, set the derivative equal to 0, and solve for c, then it follows that the value of c that minimizes $\text{Var}(Y + c(V - \mu_V))$ is

$$c^* = -\frac{\text{Cov}(Y, V)}{\text{Var}(V)}.$$

Substituting this value back into Equation (9.1) yields

$$\text{Var}(Y + c^*(V - \mu_V)) = \text{Var}(Y) - \frac{\text{Cov}^2(Y, V)}{\text{Var}(V)}. \quad (9.2)$$

Dividing both sides of this equation by $\text{Var}(Y)$ shows that

$$\frac{\text{Var}(Y + c^*(V - \mu_V))}{\text{Var}(Y)} = 1 - \text{Corr}^2(Y, V),$$

where

$$\text{Corr}(Y, V) = \frac{\text{Cov}(Y, V)}{\sqrt{\text{Var}(Y)\,\text{Var}(V)}}$$

is the correlation between Y and V. Hence, the variance reduction obtained when using the *control variate* V is $100\,\text{Corr}^2(Y, V)$ percent.

The quantities $\mathrm{Cov}(Y, V)$ and $\mathrm{Var}(V)$, which are needed to determine c^*, are not usually known and must be estimated from the simulated data. If k simulation runs produce the output Y_i and V_i ($i = 1, \ldots, k$) then, letting

$$\bar{Y} = \sum_{i=1}^{k} \frac{Y_i}{k} \quad \text{and} \quad \bar{V} = \sum_{i=1}^{k} \frac{V_i}{k}$$

be the sample means, $\mathrm{Cov}(Y, V)$ is estimated by

$$\frac{\sum_{i=1}^{k}(Y_i - \bar{Y})(V_i - \bar{V})}{k - 1}$$

and $\mathrm{Var}(V)$ is estimated by the sample variance

$$\frac{\sum_{i=1}^{k}(V_i - \bar{V})^2}{k - 1}.$$

Combining the preceding estimators gives the estimator of c^*, namely,

$$\widehat{c^*} = -\frac{\sum_{i=1}^{k}(Y_i - \bar{Y})(V_i - \bar{V})}{\sum_{i=1}^{k}(V_i - \bar{V})^2},$$

and produces the following controlled simulation estimator of θ:

$$\frac{1}{k} \sum_{i=1}^{k}(Y_i + \widehat{c^*}(V_i - \mu_V)).$$

Let us now see how control variables can be gainfully employed when simulating Asian option valuations. Suppose first that the present value of the final payoff is

$$Y = e^{-rn/N}\left(\sum_{i=1}^{n} \frac{S_d(i)}{n} - K\right)^{+}.$$

It is clear that Y is strongly positively correlated with

$$V = \sum_{i=0}^{n} S_d(i),$$

so one possibility is to use V as a control variable. Toward this end, we must first determine $E[V]$ as follows. Because

$$E[S_d(i)] = e^{ri/N}S(0)$$

for a risk-neutral valuation, we see that

$$E[V] = E\left[\sum_{i=0}^{n} S_d(i)\right]$$

$$= \sum_{i=0}^{n} E[S_d(i)]$$

$$= S(0)\sum_{i=0}^{n}(e^{r/N})^i$$

$$= S(0)\frac{1 - e^{r(n+1)/N}}{1 - e^{r/N}}.$$

Another choice of control variable that could be used is the payoff from a vanilla option with the same strike price and exercise time. That is, we could let

$$V = (S_d(n) - K)^+$$

be the control variable.

A different variance reduction technique that can be effectively employed in this case is to use *antithetic variables*. This method generates the data $X(1), \ldots, X(n)$ and uses them to compute Y. However, rather than generating a second set of data, it re-uses the same data with the following changes:

$$X(i) \implies \frac{2(r - \sigma^2/2)}{N} - X(i).$$

That is, it lets the new value of $X(i)$ be $2(r - \sigma^2/2)/N$ minus its old value, for each $i = 1, \ldots, n$. (The new value of $X(i)$ will be negatively correlated with the old value, but it will still be normal with the same mean and variance.) The value of Y based on these new values is then computed, and the estimate from that simulation run is the average of the two Y values obtained. It can be shown (see [5]) that re-using the data in this manner will result in a smaller variance than would be obtained by generating a new set of data.

Now let us consider an Asian call option for which the strike price is the average end-of-day price; that is, the present value of the final payoff is

$$Y = e^{-rn/N}\left(S_d(n) - \sum_{i=1}^{n} \frac{S_d(i)}{n}\right)^+ .$$

Recall that a simulation run consists of (a) generating $X(1), \ldots, X(n)$ independent normal random variables with mean $(r - \sigma^2/2)/N$ and variance σ^2/N, and (b) setting

$$S_d(i) = S(0)e^{X(1)+\cdots+X(i)}, \quad i = 1, \ldots, n.$$

Since the value of Y will be large if the latter values of the the sequence $X(1), X(2), \ldots, X(n)$ are among the largest (and small if the reverse is true), one could try a control variable of the type

$$V = \sum_{i=1}^{n} w_i X(i),$$

where the weights w_i are increasing in i. However, we recommend that one use all of the variables $X(1), X(2), \ldots, X(n)$ as control variables. That is, from each run one should consider the estimator

$$Y + \sum_{i=1}^{n} c_i \left(X(i) - \frac{r - \sigma^2/2}{N} \right).$$

Because the control variables are independent, it is easy to verify (see Exercise 9.4) that the optimal values of the c_i are

$$c_i = -\frac{\text{Cov}(X(i), Y)}{\text{Var}(X(i))}, \quad i = 1, \ldots, n;$$

these quantities can be estimated from the output of the simulation runs. We suggest this same approach in the case of lookback options also: again, use all of the variables $X(1), X(2), \ldots, X(n)$ as control variables.

9.6.2   Combining Conditional Expectation and Importance Sampling in the Simulation of Barrier Option Valuations

In Section 9.5 we presented a simulation approach for determining the expected value of the risk-neutral payoff under geometric Brownian motion of a down-and-in barrier call option. The $X(i)$ were generated and used to calculate the successive end-of-day prices and the resulting pay-off from the option. We can improve upon this approach by noting that, in order for this option to become alive, at least one of the end-of-day prices must fall below the barrier. Suppose that with the generated data this first occurs at the end of day j, with the price at the end of that day being $S_d(j) = x < v$. At this moment the barrier option becomes alive and its worth is exactly that of an ordinary vanilla call option, given that the price of the security is x when there is time $(n - j)/N$ that remains before the option expires. But this implies that the option's worth is now $C(x, (n - j)/N, K)$. Consequently, it seems that we could (a) end the simulation run once an end-of-day price falls below the barrier, and (b) use the resulting Black–Scholes valuation as the estimator from this run. As a matter of fact, we can do this; the resulting estimator, called the *conditional expectation estimator,* can be shown to have a smaller variance than the one derived in Section 9.5.

The conditional expectation estimator can be further improved by making use of the simulation idea of *importance sampling.* Since many of the simulation runs will never have an end-of-day price fall below the barrier, it would be nice if we could first simulate the data from a set of probabilities that makes it more likely for an end-of-day price to fall below the barrier and then add a factor to compensate for these different probabilities. This is exactly what importance sampling does. It generates the random variables $X(1), X(2), \ldots$ from a normal distribution with mean $(r - \sigma^2/2)/N - b$ and variance σ^2/N, and it determines the first time that a resulting end-of-day price falls below the barrier. If the price first falls below the barrier at time j with price x, then the estimator from that run is

$$C(x, (n - j)/N, K) \exp\left\{ \frac{jb^2 N}{2\sigma^2} + \frac{Nb}{\sigma^2} \sum_{i=1}^{j} X_i - \frac{jb}{\sigma^2}\left(r - \frac{\sigma^2}{2} \right) \right\}$$

(see [6] for details); if the price never falls below the barrier then the estimator from that run is 0. The average of these estimators over many runs is the overall estimator of the value of the option. Of course, in order to implement this procedure one needs an appropriate choice of b. Probably the best approach to choosing b is empirical; do some small simulations in cases of interest, and see which value of b leads to a small variance. In addition, the choice

$$b = \frac{r - \sigma^2/2}{N} - \frac{2\log\left(\frac{S(0)}{v}\right) + \log\left(\frac{K}{S(0)}\right)}{n}$$

was shown (in [1]) to work well for a less efficient variation of our method.

9.7   Options with Nonlinear Payoffs

The standard call option has a payoff that, provided the security's price at exercise time is in the money, is a linear function of that price. However, there are more general options whose payoff is of the form

$$\left(h(S(t)) - K\right)^+,$$

where h is an arbitrary specified function, t is the exercise time, and K is the strike price. Whereas a simulation or a numerical procedure based on a multiperiod binomial approximation to geometric Brownian motion is often needed to determine the geometric Brownian motion risk-neutral valuations of these options, an exact formula can be derived when h is of the form

$$h(x) = x^\alpha.$$

Options having nonlinear payoffs $(S^\alpha(t) - K)^+$ are called *power options,* and α is called the power parameter.

Let $C_\alpha(s, t, K, \sigma, r)$ be the risk-neutral valuation of a power call option with power parameter α that expires at time t with an exercise price K, when the interest rate is r, the underlying security initially has price s, and the security follows a geometric Brownian motion with volatility σ. As usual, let $C(s, t, K, \sigma, r) = C_1(s, t, K, \sigma, r)$ be the Black–Scholes valuation. Also, let X be a normal random variable with mean $(r - \sigma^2/2)t$ and variance $\sigma^2 t$. Because e^X has the same probability distribution as does $S(t)/s$, it follows that

$$e^{rt}C(s, t, K, \sigma, r) = E[(S(t) - K)^+] = E[(se^X - K)^+]. \qquad (9.3)$$

In addition, since $(S(t)/s)^\alpha = S^\alpha(t)/s^\alpha$ has the same distribution as does $e^{\alpha X}$, it follows that

$$E[(S^\alpha(t) - K)^+] = E[(s^\alpha e^{\alpha X} - K)^+]. \qquad (9.4)$$

But since αX is a normal random variable with mean $\alpha(r - \sigma^2/2)t$ and variance $\alpha^2\sigma^2 t$, it follows from Equation (9.3) that if we let r_α and σ_α be such that

$$r_\alpha - \sigma_\alpha^2/2 = \alpha(r - \sigma^2/2) \quad \text{and} \quad \sigma_\alpha^2 = \alpha^2\sigma^2$$

then

$$e^{r_\alpha t}C(s^\alpha, t, K, \sigma_\alpha, r_\alpha) = E[(s^\alpha e^{\alpha X} - K)^+].$$

Hence, from Equation (9.4) we obtain that

$$e^{-rt}E[(S^\alpha(t) - K)^+]$$
$$= e^{-rt}e^{r_\alpha t}C(s^\alpha, t, K, \alpha\sigma, r_\alpha)$$
$$= \exp\{(\alpha(r - \sigma^2/2) + \alpha^2\sigma^2/2 - r)t\}C(s^\alpha, t, K, \alpha\sigma, r_\alpha)$$
$$= \exp\{(\alpha - 1)(r + \alpha\sigma^2/2)t\}C(s^\alpha, t, K, \alpha\sigma, r_\alpha).$$

That is,

$$C_\alpha(s, t, K, \sigma, r) = \exp\{(\alpha - 1)(r + \alpha\sigma^2/2)t\}C(s^\alpha, t, K, \alpha\sigma, r_\alpha),$$

where

$$r_\alpha = \alpha(r - \sigma^2/2) + \alpha^2\sigma^2/2.$$

9.8 Pricing Approximations via Multiperiod Binomial Models

Multiperiod binomial models can also be used to determine efficiently the risk-neutral geometric Brownian motion prices of certain exotic options. For instance, consider the down-and-out barrier call option having initial price s, strike price K, exercise time $t = n/N$ (where N is the number of trading days in a year), and barrier value v ($v < s$). To begin, choose an integer j, let $m = nj$, and let $t_k = kt/m$ ($k = 0, 1, \ldots, m$). We will consider each day as consisting of j periods and willl approximate using an m-period binomial model that supposes

$$S(t_{k+1}) = \begin{cases} uS(t_k) & \text{with probability } p, \\ dS(t_k) & \text{with probability } 1 - p, \end{cases}$$

where

$$u = e^{\sigma\sqrt{t/m}}, \qquad d = e^{-\sigma\sqrt{t/m}},$$

$$p = \frac{1 + rt/m - d}{u - d}.$$

As in Section 7.4, we note that if i of the first k price movements are increases and $k - i$ are decreases, then the price at time t_k is

$$S(t_k) = u^i d^{k-i} s.$$

Letting $V_k(i)$ denote the expected payoff from the barrier call option given that the option is still alive at time t_k and that the price at time t_k is $S(t_k) = u^i d^{k-i} s$, we can approximate the expected present value payoff of the European barrier call option by $e^{-rt} V_0(0)$. The value of $V_0(0)$ can be obtained by working backwards. That is, we start with the identity

$$V_m(i) = (u^i d^{m-i} s - K)^+, \quad i = 0, \ldots, m,$$

to determine the values of $V_m(i)$ and then repeatedly use the following equation (initially with $k = m - 1$, and then decreasing its value by 1 after each interation):

$$V_k(i) = pV_{k+1}(i + 1) + (1 - p)W_{k+1}(i), \qquad (9.5)$$

where

$$W_{k+1}(i) = \begin{cases} 0 & \text{if } u^i d^{k+1-i} s < v \text{ and } j \text{ divides } k + 1, \\ V_{k+1}(i) & \text{otherwise.} \end{cases}$$

Note that $W_{k+1}(i)$ is defined in this fashion because if j divides $k + 1$ then the period-$(k + 1)$ price is an end-of-day price and will thus kill the option if it is less than the barrier value.

If we wanted the risk-neutral price of a down-and-in call option then we could use an analogous procedure. Alternatively, we could use the preceding to determine the price of a down-and-out call option with the same parameters and then use the identity

$$D_i(s, t, K) + D_o(s, t, K) = C(s, t, K),$$

where D_i, D_o, and C refer to the risk-neutral price of (respectively) a down-and-in call option, a down-and-out call option, and a vanilla Black–Scholes call option.

Risk-neutral prices of other exotic options can also be approximated by multiperiod binomial models. However, the computational burden can be demanding. For instance, consider an Asian option whose strike price is the average of the end-of-day prices. To recursively determine the expected value of the final payoff given all that has occurred up to time t_k, we need to specify not only the price at time t_k but also the sum of the end-of-day prices up to that time. That is, in order to approximate an n-day call option with an n-period binomial model, we would need to recursively compute the values $V_k(i, x)$ equal to the expected final payoff given that the price after k periods is $u^i d^{k-i} s$ and that the sum of the first k prices is x. Since there can be as many as $\binom{k}{i}$ possible sums of the first k prices when i of them are increases, it can require a great deal of computation to obtain a good approximation. Generally speaking, we recommend the use of simulation to estimate the risk-neutral prices of most path-dependent exotic options.

9.9 Exercises

Exercise 9.1 Consider an American call option that can be exercised at any time up to time t; however, if it is exercised at time y (where $0 \leq y \leq t$) then the strike price is Ke^{uy} for some specified value of u. That is, the payoff if the call is exercised at time y ($0 \leq y \leq t$) is

$$(S(y) - e^{uy}K)^+.$$

Argue that if $u \leq r$ then the call should never be exercised early, where r is the interest rate.

Exercise 9.2 A lookback put option that expires after n trading days has a payoff equal to the maximum end-of-day price achieved by time n minus the price at time n. That is, the payoff is

$$\max_{0 \leq i \leq n} S_d(i) - S_d(n).$$

Explain how Monte Carlo simulation can be used efficiently to find the geometric Brownian motion risk-neutral price of such an option.

Exercise 9.3 In Section 9.6.1, it is noted that $V = (S_d(n) - K)^+$ can be used as a control variate. However, doing so requires that we know its mean; what is $E[V]$?

Exercise 9.4   Let $X_1, \ldots, X_n$ be independent random variables with expected values $E[X_i] = \mu_i$, and consider the following simulation estimator of $E[Y]$:

$$W = Y + \sum_{i=1}^{n} c_i (X_i - \mu_i).$$

(a) Show that

$$\mathrm{Var}(W) = \mathrm{Var}(Y) + \sum_{i=1}^{n} c_i^2 \, \mathrm{Var}(X_i) + 2 \sum_{i=1}^{n} c_i \, \mathrm{Cov}(Y, X_i).$$

(b) Use calculus to show that the values of $c_1, \ldots, c_n$ that minimize $\mathrm{Var}(W)$ are

$$c_i = -\frac{\mathrm{Cov}(Y, X_i)}{\mathrm{Var}(X_i)}, \quad i = 1, \ldots, n.$$

Exercise 9.5   Perform a Monte Carlo simulation to estimate the risk-neutral valuation of some exotic option. Do it first without any attempts at variance reduction and then a second time with some variance reduction procedure.

Exercise 9.6   Give the equations that are needed when using a multi-period binomial model to approximate the risk-neutral price of a down-and-in barrier call option.

Exercise 9.7   Explain how you can approximate the risk-neutral price of a down-and-out *American* call option by using a multiperiod binomial model.

Exercise 9.8   Explain why Equation (9.5) is valid.

REFERENCES

[1] Boyle, P., M. Broadie, and P. Glasserman (1997). "Monte Carlo Methods for Security Pricing." *Journal of Economic Dynamics and Control* 21: 1267–1321.

[2] Conze, A., and R. Viswanathan (1991). "Path Dependent Options: The Case of Lookback Options." *Journal of Finance* 46: 1893–1907.

[3] Goldman, B., H. Sosin, and M. A. Gatto (1979). "Path Dependent Options: Buy at the Low, Sell at the High." *Journal of Finance* 34: 1111–27.

[4] Hull, J. C., and A. White (1998). "The Use of the Control Variate Technique in Option Pricing." *Journal of Financial and Quantitative Analysis* 23: 237–51.

[5] Ross, S. M. (1997). *Simulation,* 2nd ed. Orlando, FL: Academic Press.

[6] Ross, S. M., and J. G. Shanthikumar (1999). "Pricing Exotic Options: Monotonicity in Volatility and Efficient Simulations." Unpublished manuscript.

[7] Rubinstein, M. (1991). "Pay Now, Choose Later." *Risk* (February).

10. Beyond Geometric Brownian Motion Models

10.1   Introduction

As previously noted, a key premise underlying the assumption that the prices of a security over time follow a geometric Brownian motion (and hence underlying the Black–Scholes option price formula) is that future price changes are independent of past price movements. Many investors would agree with this premise, although many others would disagree. Those accepting the premise might argue that it is a consequence of the *efficient market hypothesis,* which claims that the present price of a security encompasses all the presently available information – including past prices – concerning this security. However, critics of this hypothesis argue that new information is absorbed by different investors at different rates; thus, past price movements are a reflection of information that has not yet been universally recognized but *will* affect future prices. It is our belief that there is no a priori reason why future price movements should necessarily be independent of past movements; one should therefore look at real data to see if they are consistent with the geometric Brownian motion model. That is, rather than taking an a priori position, one should let the data decide as much as possible.

In Section 10.2 we analyze the sequence of nearest-month end-of-day prices of crude oil from 3 January 1995 to 19 November 1997 (a period right before the beginning of the Asian financial crisis that deeply affected demand and, as a result, led to lower crude prices). As part of our analysis, we argue that such a price sequence is not consistent with the assumption that crude prices follow a geometric Brownian motion. In Section 10.3 we offer a new model that is consistent with the data as well as intuitively plausible, and we indicate how it may be used to obtain option prices under (a) the assumption that the future resembles the past and (b) a risk-neutral valuation based on the new model.

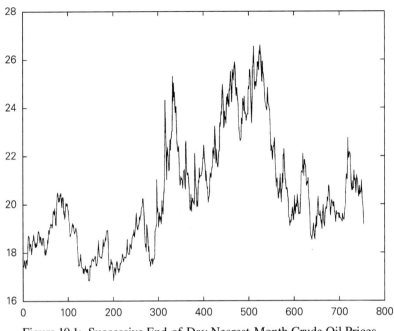

Figure 10.1: Successive End-of-Day Nearest-Month Crude Oil Prices

10.2 Crude Oil Data

With day 0 defined to be 3 January 1995, let $P(n)$ denote the nearest-month price of crude oil (as traded on the New York Mercantile Exchange) at the end of the nth trading day from day 0. The values of $P(n)$ for $n = 1, \ldots, 752$ are given in Figure 10.1 (and in Table 10.5, located at the end of this chapter).

Let

$$L(n) = \log(P(n)),$$

and define

$$D(n) = L(n) - L(n - 1).$$

That is, $D(n)$ for $n \geq 1$ are the successive differences in the logarithms of the end-of-day prices. The values of the $D(n)$ are also given in Table 10.5, and Figure 10.2 presents a histogram of those data.

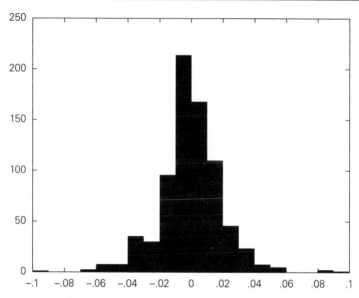

Figure 10.2: Histogram of Log Differences

Note that, under geometric Brownian motion, the $D(n)$ would be independent and identically distributed normal random variables; the histogram in Figure 10.2 is consistent with the hypothesis that the data come from a normal population. However, a histogram – which breaks up the range of data values into intervals and then plots the number of data values that fall in each interval – is not informative about possible dependencies among the data. To consider this possibility, let us classify each day as being in one of four possible *states* as follows: the state of day n is

$$1 \quad \text{if } D(n) \le -.01,$$

$$2 \quad \text{if } -.01 < D(n) \le 0,$$

$$3 \quad \text{if } 0 < D(n) \le .01,$$

$$4 \quad \text{if } D(n) > .01.$$

That is, day n is in state 1 if its end-of-day price represents a loss of more than 1% ($e^{-.01} \approx .99005$) from the end-of-day price on day $n-1$;

Table 10.1

		j			
i	1	2	3	4	Total
1	55	41	44	36	176
2	44	65	45	60	214
3	26	46	47	49	168
4	52	62	31	48	193

it is in state 2 if the percentage loss is less than 1%; it is in state 3 if the percentage *gain* is less than 1% ($e^{.01} \approx 1.0101$); and it is in state 4 if its end-of-day price represents a gain of more than 1% from the end-of-day price on day $n - 1$. Note that, if the price evolution follows a geometric Brownian motion, then tomorrow's state will not depend on today's state. One way to verify the plausibility of this hypothesis is to see how many times that a state i day was followed by a state j day for $i, j = 1, \ldots, 4$. Table 10.1 gives this information and shows, for instance, that 26 of the 168 days in state 3 were followed by a state-1 day, 46 were followed by a state-2 day, and so on.

The implications of Table 10.1 become clearer if we express the data in terms of percentages, as is done in Table 10.2. Thus, for instance, a large drop (more than 1%) was followed 31% of the time by another large drop, 23% of the time by a small drop, 25% of the time by a small increase, and 21% of the time by a large increase. It is interesting to note that, whereas a moderate gain was followed by a large drop 15% of the time, a large gain was followed by a large drop 27% of the time. Under the geometric Brownian motion model, tomorrow's change would be unaffected by today's change and so the theoretically expected percentages in Table 10.2 would be the same for all rows. To see how likely it is that the actual data would have occurred under geometric Brownian motion, we can employ a standard statistical procedure (testing for independence in a contingency table); using this procedure on our data results in a p-value equal to .005. This means that if the row probabilities were equal (as implied by geometric Brownian motion), then the probability that the resulting data would be as nonsupportive of this hypothesized equality as our actual data is only about 1 in

Table 10.2

		j		
i	1	2	3	4
1	31	23	25	21
2	21	30	21	28
3	15	28	28	29
4	27	32	16	25

200. (The value of the test statistics is 23.447, resulting in a p-value of .00526.)

Let us now break up the data, which consists of 751 $D(n)$ values, into four groupings: the first group consists of the 176 values (of the log of tomorrow's price minus the log of today's) for which today's state is 1, and so on with the other groupings. Figures 10.3–10.6 present the histograms of the data values in each group. Note that each histogram has (approximately) the bell-shaped form of the normal density function.

Let $\bar{x}_i$ and s_i be, respectively, the sample mean and sample standard deviation (equal to the square root of the sample variance) of grouping i for $i = 1, 2, 3, 4$. A computation produces the values listed in Table 10.3.

Under the geometric Brownian motion model, the four data sets will all come from the same normal population and hence we could use a standard statistical test – called a one-way analysis of variance – to test the hypothesis that all four data sets describe normal random variables having the same mean and variance. The necessary calculations reveal that the test statistic (which, when the hypothesis is true, has an F distribution with 3 numerator and 747 denominator degrees of freedom) has a value of 4.50, which is quite large. Indeed, if the hypothesis were true then the probability that the test statistic would have a value at least this large is less than .001, giving us additional evidence that the crude oil data does not follow a geometric Brownian motion. (We could also test the hypothesis that the variances – but not necessarily the means – are equal by using Bartlett's test for the equality of variances; using our data, the test statistic has value 9.59 with a resulting p-value less than .025.)

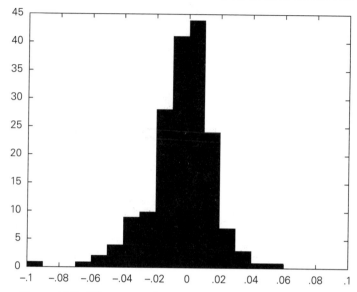

Figure 10.3: Histogram of Post–State-1 Outcomes ($n = 176$)

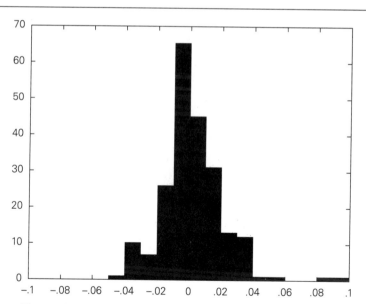

Figure 10.4: Histogram of Post–State-2 Outcomes ($n = 214$)

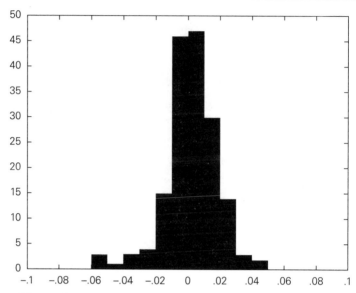

Figure 10.5: Histogram of Post–State-3 Outcomes ($n = 168$)

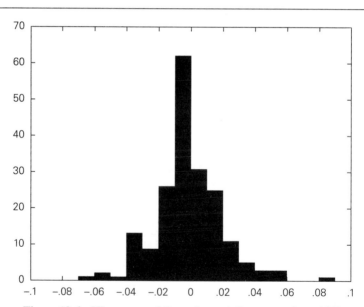

Figure 10.6: Histogram of Post–State-4 Outcomes ($n = 193$)

Table 10.3

i	Mean $\bar{x}_i$	S.D. s_i
1	−.0036	.0194
2	.0024	.0188
3	.0025	.0165
4	−.0011	.0208

10.3   Models for the Crude Oil Data

A reasonable model is to suppose that there are four distributions that
determine the difference between the logarithm of tomorrow's price and
the logarithm of today's, with the appropriate distribution depending on
today's state. However, even within this context we still need to decide
if we want a risk-neutral model or one based on the assumption that
the future will tend to follow the past. In the latter case we could use a
model that supposes, if today's state is i, that the logarithm of the ratio
of tomorrow's price to today's price is a normal random variable with
mean $\bar{x}_i$ and standard deviation s_i, where these quantities are as given in
Table 10.3. However, it is quite possible that a better model is obtained
by forgoing the normality assumption and using instead a "bootstrap"
approach, which supposes that the best approximation to the distribution of a log ratio from state i is obtained by randomly choosing one
of the n_i data values in this grouping (where, in the present situation,
$n_1 = 176$, $n_2 = 214$, $n_3 = 168$, and $n_4 = 193$). Whether we assume
that the group data are normal or instead use a bootstrap approach, a
Monte Carlo simulation (see Chapter 9) will be needed to determine
the expected value of owning an option – or even the expected value
of a future price. However, such a simulation is straightforward, and
variance reduction techniques are available that can reduce the computational time.

A risk-neutral model would appear to be the most appropriate type
for assessing whether a specified option is underpriced or overpriced in
relation to the present price of the security. Such a model is obtained in
the present situation by supposing that, when in state i, the next log ratio is a normal random variable with standard deviation (i.e. volatility)

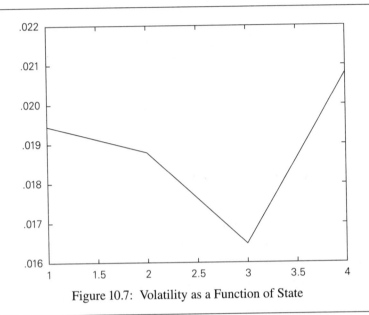

Figure 10.7: Volatility as a Function of State

s_i and mean μ_i, where

$$\mu_i = r/N - s_i^2/2;$$

r is the interest rate, and N (usually taken equal to 252) is the number of trading days in a year. Again, a simulation would be needed to determine the expected worth of an option.

Whereas we have chosen to define four different states depending on the ratio of successive end-of-day prices, it is quite possible that a better model could be obtained by allowing for more states. Indeed, one approach for obtaining a risk-neutral model is to assess the volatility as a function of the most recent value of $D(n)$ – by assuming that the volatility is equal to s_i when $D(n)$ is the midpoint of region i – and then to use a general linear interpolation scheme (see Figure 10.7).

Rather than having four different states, we might rather have defined *six* states as follows: the state of day n is

1   if $D(n) \leq -.02$,

2   if $-.02 < D(n) \leq .01$,

Table 10.4

			j				
i	1	2	3	4	5	6	Total
1	10	12	25	19	12	3	81
2	17	16	16	25	12	9	95
3	18	26	65	45	31	29	214
4	11	15	46	47	30	19	168
5	14	15	39	19	13	10	110
6	12	11	23	12	12	13	83

$$3 \quad \text{if} \quad -.01 < D(n) \leq 0,$$

$$4 \quad \text{if} \quad 0 < D(n) \leq .01,$$

$$5 \quad \text{if} \quad .01 < D(n) \leq .02,$$

$$6 \quad \text{if} \quad D(n) > .02.$$

With these states, the number of times that a state-i day was followed by a state-j day is as given in row i, column j of Table 10.4. The resulting model can then be analyzed in exactly the same manner as was the four-state model.

10.4 Final Comments

We have seen in this chapter that not all security price data is consistent with the assumption that its price history follows a geometric Brownian motion. Geometric Brownian motion is a Markov model, which is one that supposes that a future state of the system (i.e., price of the security) depends only on the present state and not on any previous states. However, to many people it seems reasonable that a security's recent price history can be somewhat useful in predicting future prices. In this chapter we have proposed a simple model for end-of-day prices, one in which the successive ratios of the price on day n to the price on day $n-1$ are assumed to constitute a Markov model. That is, with regard to the successive ratios of prices, geometric Brownian motion supposes that they are independent whereas our proposed model allows them to have a Markov dependence.

In using the model to value an option, we recommend that one collect up-to-date data and then model the future under the assumption that it will follow the past, either by using a bootstrap approach or by assuming normality and using the estimates $\bar{x}_i$ and s_i. However, if one wants to determine whether an option is underpriced or overpriced in relation to the security itself, we recommend using the risk-neutral variant of the model. This latter model takes $r/N - s_i^2/2$, rather than $\bar{x}_i$, as the mean of a log ratio from state i. This risk-neutral model, which allows the volatility to depend on the most recent daily change, is consistent with a variant of the efficient market hypothesis which states that the present price of a security is the "fair price," in the sense that the expectation of the present value of a future price is equal to the present price (this is known as the *martingale* hypothesis).

REFERENCES

[1] Efron, B., and R. Tibshirani (1993). *An Introduction to the Bootstrap.* New York: Chapman and Hall.
[2] Fama, Eugene (1965). "The Behavior of Stock Market Prices." *Journal of Business* 38: 34–105.
[3] Malkiel, Burton G. (1990). *A Random Walk Down Wall Street.* New York: Norton.
[4] Niederhoffer, Victor (1966). "A New Look at Clustering of Stock Prices." *Journal of Business* 39: 309–13.

Table 10.5: *Nearest-Month Crude Oil Data* (dollars)

Date	Price	Log Difference	Date	Price	Log Difference
1/3/95	17.44		3/7/95	18.63	0.00214938
1/4/95	17.48	0.00229095	3/8/95	18.33	−0.0162341
1/5/95	17.72	0.0136366	3/9/95	18.02	−0.0170568
1/6/95	17.67	−0.00282566	3/10/95	17.91	−0.00612304
1/9/95	17.4	−0.0153981	3/13/95	18.19	0.0155128
1/10/95	17.37	−0.00172563	3/14/95	17.94	−0.0138391
1/11/95	17.72	0.0199494	3/15/95	18.11	0.00943142
1/12/95	17.72	0	3/16/95	18.16	0.0027571
1/13/95	17.52	−0.0113509	3/17/95	18.26	0.0054915
1/16/95	17.88	0.0203397	3/20/95	18.56	0.0162959
1/17/95	18.32	0.0243106	3/21/95	18.43	−0.00702896
1/18/95	18.73	0.0221332	3/22/95	18.96	0.0283517
1/19/95	18.69	−0.00213789	3/23/95	18.92	−0.00211193
1/20/95	18.65	−0.00214248	3/24/95	18.78	−0.00742709
1/23/95	18.1	−0.0299342	3/27/95	19.07	0.0153239
1/24/95	18.39	0.0158951	3/28/95	19.05	−0.00104932
1/25/95	18.39	0	3/29/95	19.22	0.0088843
1/26/95	18.24	−0.00819005	3/30/95	19.15	−0.00364869
1/27/95	17.95	−0.0160269	3/31/95	19.17	0.00104384
1/30/95	18.09	0.00776918	4/3/95	19.03	−0.00732988
1/31/95	18.39	0.0164477	4/4/95	19.18	0.00785139
2/1/95	18.52	0.00704419	4/5/95	19.56	0.0196186
2/2/95	18.54	0.00107933	4/6/95	19.77	0.010679
2/3/95	18.78	0.0128619	4/7/95	19.67	−0.005071
2/6/95	18.59	−0.0101687	4/10/95	19.59	−0.0040754
2/7/95	18.46	−0.00701757	4/11/95	19.88	0.014695
2/8/95	18.3	−0.00870517	4/12/95	19.55	−0.0167389
2/9/95	18.24	−0.00328408	4/13/95	19.15	−0.0206726
2/10/95	18.46	0.0119892	4/17/95	19.15	0
2/13/95	18.27	−0.0103459	4/18/95	19.73	0.0298376
2/14/95	18.32	0.00273299	4/19/95	20.05	0.0160888
2/15/95	18.42	0.00544367	4/20/95	20.41	0.0177958
2/16/95	18.59	0.00918677	4/21/95	20.52	0.00537504
2/17/95	18.91	0.0170671	4/24/95	20.41	−0.00537504
	18.91	0	4/25/95	20.12	−0.0143106
2/21/95	18.86	−0.00264761	4/26/95	20.29	0.00841381
2/22/95	18.63	−0.0122701	4/27/95	20.15	−0.00692387
2/23/95	18.43	−0.0107934	4/28/95	20.43	0.0138001
2/24/95	18.69	0.0140088		20.38	−0.00245038
2/27/95	18.66	−0.00160643	5/1/95	20.5	0.00587086
2/28/95	18.49	−0.00915215	5/2/95	20.09	−0.0202027
3/1/95	18.32	−0.00923669	5/3/95	19.89	−0.0100051
3/2/95	18.35	0.00163622	5/4/95	20.29	0.0199111
3/3/95	18.63	0.0151436	5/5/95	20.33	0.00196947
3/6/95	18.59	−0.00214938	5/8/95	20.29	−0.00196947

Table 10.5 *(cont.)*

Date	Price	Log Difference	Date	Price	Log Difference
5/9/95	19.61	−0.0340885	7/11/95	17.32	−0.00115407
5/10/95	19.75	0.00711385	7/12/95	17.49	0.00976739
5/11/95	19.41	−0.0173651	7/13/95	17.25	−0.0138171
5/12/95	19.52	0.00565118	7/14/95	17.32	0.00404976
5/15/95	19.9	0.0192802	7/17/95	17.2	−0.00695252
5/16/95	20.08	0.00900456	7/18/95	17.35	0.00868312
5/17/95	19.96	−0.00599402	7/19/95	17.33	−0.0011534
5/18/95	20	0.002002	7/20/95	17.01	−0.0186377
5/19/95	20.06	0.00299551	7/21/95	16.79	−0.0130179
5/22/95	19.81	−0.0125409	7/24/95	16.88	0.00534602
5/23/95	19.77	−0.00202122	7/25/95	16.93	0.00295771
5/24/95	19.41	−0.0183772	7/26/95	17.5	0.0331137
5/25/95	19.26	−0.00775799	7/27/95	17.49	−0.000571592
5/26/95	18.69	−0.0300418	7/28/95	17.43	−0.00343643
	18.69	0	7/31/95	17.56	0.00743073
5/30/95	18.78	0.00480385	8/1/95	17.7	0.00794105
5/31/95	18.89	0.00584021	8/2/95	17.78	0.00450959
6/1/95	18.9	0.000529241	8/3/95	17.72	−0.00338028
6/2/95	19.14	0.0126185	8/4/95	17.71	−0.000564493
6/5/95	19.25	0.00573067	8/7/95	17.65	−0.00339367
6/6/95	19.06	−0.00991916	8/8/95	17.79	0.00790072
6/7/95	19.18	0.00627617	8/9/95	17.78	−0.000562272
6/8/95	18.91	−0.0141772	8/10/95	17.89	0.00616767
6/9/95	18.8	−0.00583401	8/11/95	17.86	−0.00167832
6/12/95	18.86	0.00318641	8/14/95	17.48	−0.0215062
6/13/95	18.91	0.00264761	8/15/95	17.47	−0.000572246
6/14/95	19.05	0.00737622	8/16/95	17.55	0.00456883
6/15/95	18.94	−0.00579101	8/17/95	17.66	0.00624825
6/16/95	18.84	−0.00529382	8/18/95	17.87	0.0118211
6/19/95	18.22	−0.0334624	8/21/95	18.25	0.0210418
6/20/95	18.01	−0.0115927	8/22/95	18.54	0.0157655
6/21/95	17.46	−0.0310146	8/23/95	18	−0.0295588
6/22/95	17.5	0.00228833	8/24/95	17.86	−0.00780818
6/23/95	17.49	−0.000571592	8/25/95	17.86	0
6/26/95	17.64	0.00853976	8/28/95	17.82	−0.00224215
6/27/95	17.77	0.00734259	8/29/95	17.82	0
6/28/95	17.97	0.0111921	8/30/95	17.79	−0.00168492
6/29/95	17.56	−0.0230801	8/31/95	17.84	0.00280663
6/30/95	17.4	−0.00915338	9/1/95	18.04	0.0111484
	17.4	0		18.04	0
	17.4	0	9/5/95	18.58	0.0294942
7/5/95	17.18	−0.0127243	9/6/95	18.36	−0.0119113
7/6/95	17.37	0.0109987	9/7/95	18.27	−0.00491401
7/7/95	17.14	−0.0133297	9/8/95	18.44	0.00926185
7/10/95	17.34	0.0116011	9/11/95	18.47	0.00162558

Table 10.5 *(cont.)*

Date	Price	Log Difference	Date	Price	Log Difference
9/12/95	18.64	0.00916202	11/14/95	17.82	0.00112296
9/13/95	18.54	−0.00537925	11/15/95	17.93	0.00615387
9/14/95	18.85	0.0165824	11/16/95	18.19	0.0143967
9/15/95	18.92	0.00370665	11/17/95	18.57	0.0206754
9/18/95	18.93	0.000528402	11/20/95	18.06	−0.0278478
9/19/95	18.95	0.00105597	11/21/95	17.97	−0.00499585
9/20/95	18.69	−0.0138153	11/22/95	17.96	−0.000556638
9/21/95	17.56	−0.062365	11/23/95	17.96	0
9/22/95	17.25	−0.0178114	11/24/95	17.96	0
9/25/95	17.47	0.012673	11/27/95	18.38	0.0231161
9/26/95	17.33	−0.00804602	11/28/95	18.33	−0.00272406
9/27/95	17.57	0.0137538	11/29/95	18.26	−0.00382619
9/28/95	17.76	0.0107558	11/30/95	18.18	−0.00439079
9/29/95	17.54	−0.0124648	12/1/95	18.43	0.0136577
10/2/95	17.64	0.00568506	12/4/95	18.63	0.0107934
10/3/95	17.56	−0.00454546	12/5/95	18.67	0.00214477
10/4/95	17.3	−0.0149171	12/6/95	18.77	0.00534189
10/5/95	16.87	−0.0251696	12/7/95	18.73	−0.00213333
10/6/95	17.03	0.0094396	12/8/95	18.97	0.0127323
10/9/95	17.31	0.0163079	12/11/95	18.66	−0.0164766
10/10/95	17.42	0.0063346	12/12/95	18.73	0.00374432
10/11/95	17.29	−0.00749067	12/13/95	19	0.0143125
10/12/95	17.12	−0.00988093	12/14/95	19.11	0.00577278
10/13/95	17.41	0.0167974	12/15/95	19.51	0.0207154
10/16/95	17.59	0.0102858	12/18/95	19.67	0.00816748
10/17/95	17.68	0.0051035	12/19/95	19.12	−0.0283597
10/18/95	17.61	−0.00396713	12/20/95	18.97	−0.00787612
10/19/95	17.32	−0.016605	12/21/95	18.96	−0.000527287
10/20/95	17.37	0.00288268	12/22/95	19.14	0.00944889
10/23/95	17.21	−0.00925397	12/25/95	19.14	0
10/24/95	17.32	0.00637129	12/26/95	19.27	0.0067691
10/25/95	17.32	0	12/27/95	19.5	0.011865
10/26/95	17.58	0.0149	12/28/95	19.36	−0.00720538
10/27/95	17.54	−0.00227791	12/29/95	19.55	0.0097662
10/30/95	17.62	0.00455063	1/1/96	19.55	0
10/31/95	17.64	0.00113443	1/2/96	19.81	0.0132116
11/1/95	17.74	0.00565293	1/3/96	19.89	0.00403023
11/2/95	17.98	0.0134381	1/4/96	19.91	0.00100503
11/3/95	17.94	−0.00222717	1/5/96	20.26	0.0174264
11/6/95	17.71	−0.0129034	1/8/96	20.26	0
11/7/95	17.65	−0.00339367	1/9/96	19.95	−0.0154194
11/8/95	17.82	0.00958564	1/10/96	19.67	−0.0141345
11/9/95	17.84	0.00112171	1/11/96	18.79	−0.0457698
11/10/95	17.83	−0.000560695	1/12/96	18.25	−0.0291597
11/13/95	17.8	−0.00168397	1/15/96	18.38	0.00709804

Table 10.5 *(cont.)*

Date	Price	Log Difference	Date	Price	Log Difference
1/16/96	18.05	−0.0181174	3/19/96	24.34	0.0449561
1/17/96	18.52	0.0257055	3/20/96	23.06	−0.0540216
1/18/96	19.18	0.0350168	3/21/96	21.05	−0.091199
1/19/96	18.94	−0.012592	3/22/96	21.95	0.0418666
1/22/96	18.62	−0.0170398	3/25/96	22.4	0.0202938
1/23/96	18.06	−0.0305367	3/26/96	22.19	−0.00941922
1/24/96	18.28	0.012108	3/27/96	21.79	−0.0181906
1/25/96	17.67	−0.0339393	3/28/96	21.41	−0.017593
1/26/96	17.73	0.00338983	3/29/96	21.47	0.00279851
1/29/96	17.45	−0.0159185	4/1/96	22.26	0.0361347
1/30/96	17.56	0.00628394	4/2/96	22.7	0.0195736
1/31/96	17.74	0.0101984	4/3/96	22.27	−0.0191244
2/1/96	17.71	−0.00169253	4/4/96	22.75	0.0213247
2/2/96	17.8	0.00506901	4/5/96	22.75	0
2/5/96	17.54	−0.0147145	4/8/96	23.03	0.0122326
2/6/96	17.69	0.00851552	4/9/96	23.06	0.0013018
2/7/96	17.74	0.00282247	4/10/96	24.21	0.0486663
2/8/96	17.76	0.00112676	4/11/96	25.34	0.0456184
2/9/96	17.78	0.00112549	4/12/96	24.29	−0.0423194
2/12/96	17.97	0.0106295	4/15/96	25.06	0.0312082
2/13/96	18.91	0.0509872	4/16/96	24.47	−0.0238251
2/14/96	18.96	0.00264061	4/17/96	24.67	0.00814005
2/15/96	19.04	0.00421053	4/18/96	23.82	−0.0350624
2/16/95	19.16	0.00628274	4/19/96	23.95	0.00544276
2/19/96	19.16	0	4/22/96	24.07	0.00499793
2/20/96	21.05	0.0940758	4/23/96	22.7	−0.0586013
2/21/96	19.71	−0.0657744	4/24/96	22.4	−0.013304
2/22/96	19.85	0.00707789	4/25/96	22.2	−0.00896867
2/23/96	19.06	−0.0406121	4/26/96	22.32	0.00539085
2/26/96	19.39	0.0171656	4/29/96	22.43	0.00491621
2/27/96	19.7	0.0158612	4/30/96	21.2	−0.0563982
2/28/96	19.29	−0.0210318	5/1/96	20.81	−0.0185675
2/29/96	19.54	0.0128768	5/2/96	20.86	0.00239981
3/1/96	19.44	−0.00513085	5/3/96	21.18	0.0152239
3/4/96	19.2	−0.0124225	5/6/96	21.04	−0.00663195
3/5/96	19.54	0.0175534	5/7/96	21.11	0.00332147
3/6/96	20.19	0.0327238	5/8/96	21	−0.00522442
3/7/96	19.81	−0.0190006	5/9/96	20.68	−0.0153554
3/8/96	19.61	−0.0101472	5/10/96	21.01	0.0158315
3/11/96	19.91	0.0151825	5/13/96	21.36	0.0165215
3/12/96	20.46	0.0272496	5/14/96	21.42	0.00280505
3/13/96	20.58	0.00584797	5/15/96	21.48	0.0027972
3/14/96	21.16	0.0277929	5/16/96	20.78	−0.0331313
3/15/96	21.99	0.0384752	5/17/96	20.64	−0.00676005
3/18/96	23.27	0.0565772	5/20/96	22.48	0.0853951

Table 10.5 *(cont.)*

Date	Price	Log Difference	Date	Price	Log Difference
5/21/96	22.65	0.00753383	7/23/96	21.01	−0.0183924
5/22/96	21.4	−0.0567689	7/24/96	20.68	−0.0158315
5/23/96	21.23	−0.00797565	7/25/96	20.74	0.00289715
5/24/96	21.32	0.00423032	7/26/96	20.11	−0.030847
	21.32	0	7/29/86	20.28	0.00841797
5/28/96	21.11	−0.00989874	7/30/96	20.33	0.00246245
5/29/96	20.76	−0.0167188	7/31/96	20.42	0.00441719
5/30/96	19.94	−0.0403003	8/1/96	21.04	0.0299106
5/31/96	19.76	−0.00906807	8/2/96	21.34	0.0141579
6/3/96	19.85	0.00454431	8/5/96	21.23	−0.00516797
6/4/96	20.44	0.0292898	8/6/96	21.13	−0.00472144
6/5/96	19.72	−0.0358604	8/7/96	21.42	0.0136312
6/6/96	20.05	0.0165958	8/8/96	21.55	0.00605075
6/7/96	20.28	0.011406	8/9/96	21.57	0.000927644
6/10/96	20.25	−0.00148039	8/12/96	22.22	0.0296893
6/11/96	20.1	−0.00743498	8/13/96	22.37	0.00672799
6/12/96	20.09	−0.000497636	8/14/96	22.12	−0.0112386
6/13/96	20.01	−0.00399003	8/15/96	21.9	−0.00999554
6/14/96	20.34	0.0163572	8/16/96	22.66	0.0341146
6/17/96	22.14	0.0847965	8/19/96	23.26	0.0261339
6/18/96	21.46	−0.0311952	8/20/96	22.86	−0.0173465
6/19/96	20.76	−0.0331627	8/21/96	21.72	−0.0511552
6/20/96	20.65	−0.00531274	8/22/96	22.3	0.0263532
6/21/96	19.92	−0.0359911	8/23/96	21.96	−0.0153641
6/24/96	19.98	0.00300752	8/26/96	21.62	−0.0156038
6/25/96	19.96	−0.0010015	8/27/96	21.56	−0.00277907
6/26/96	20.65	0.033985	8/28/96	21.71	0.00693324
6/27/96	21.02	0.017759	8/29/96	22.15	0.0200645
6/28/96	20.92	−0.00476873	8/30/96	22.25	0.00450451
7/1/96	21.53	0.0287417	9/2/96	22.25	0
7/2/96	21.13	−0.0187535	9/3/96	23.4	0.050394
7/3/96	21.21	0.00377894	9/4/96	23.24	−0.00686109
7/4/96	21.21	0	9/5/96	23.44	0.00856903
7/5/96	21.21	0	9/6/96	23.85	0.0173403
7/8/96	21.27	0.00282486	9/9/96	23.73	−0.00504415
7/9/96	21.41	0.00656047	9/10/96	24.12	0.0163013
7/10/96	21.55	0.00651771	9/11/96	24.75	0.0257841
7/11/96	21.95	0.0183913	9/12/96	25	0.0100503
7/12/96	21.9	−0.0022805	9/13/96	24.51	−0.0197946
7/15/96	22.48	0.0261394	9/16/96	23.19	−0.05536
7/16/96	22.38	−0.00445832	9/17/96	23.31	0.0051613
7/17/96	21.8	−0.0262577	9/18/96	23.89	0.0245775
7/18/96	21.68	−0.00551979	9/19/96	23.54	−0.0147589
7/19/96	21	−0.0318677	9/20/96	23.63	0.00381599
7/22/96	21.4	0.0188685	9/23/96	23.37	−0.0110639

Table 10.5 *(cont.)*

Date	Price	Log Difference	Date	Price	Log Difference
9/24/96	24.07	0.0295131	11/26/96	23.62	0.00551901
9/25/96	24.46	0.0160729	11/27/96	23.75	0.00548872
9/26/96	24.16	−0.0123408	11/28/96	23.75	0
9/27/96	24.6	0.0180481	11/29/96	23.75	0
9/30/96	24.38	−0.00898322	12/2/96	24.8	0.0432611
10/1/96	24.14	−0.00989291	12/3/96	24.93	0.00522824
10/2/96	24.05	−0.00373522	12/4/96	24.8	−0.00522824
10/3/96	24.81	0.0311118	12/5/96	25.58	0.0309671
10/4/96	24.73	−0.00322972		25.62	0.0015625
10/7/96	25.24	0.020413	12/9/96	25.3	−0.0125689
10/8/96	25.54	0.0118158	12/10/96	24.42	−0.0354019
10/9/96	25.07	−0.0185739	12/11/96	23.38	−0.0435215
10/10/96	24.26	−0.032843	12/12/96	23.72	0.0144376
10/11/96	24.66	0.0163536	12/13/96	24.47	0.0311293
10/14/96	25.62	0.0381908	12/16/96	25.74	0.0505983
10/15/98	25.42	−0.00783703	12/17/96	25.71	−0.00116618
10/16/96	25.17	−0.00988346	12/18/96	26.16	0.0173515
10/17/96	25.42	0.00988346	12/19/96	26.57	0.0155512
10/18/96	25.75	0.0128984	12/20/96	25.08	−0.057712
10/21/96	25.92	0.00658024	12/23/96	24.79	−0.0116304
10/22/96	25.75	−0.00658024	12/24/96	25.1	0.0124275
10/23/96	24.86	−0.0351745	12/25/96	25.1	0
10/24/96	24.51	−0.0141789	12/26/96	24.92	−0.00719715
10/25/96	24.86	0.0141789	12/27/96	25.22	0.0119666
10/28/96	24.85	−0.000402334	12/30/96	25.37	0.00593004
10/29/96	24.34	−0.0207367	12/31/96	25.92	0.0214475
10/30/96	24.28	−0.00246812		25.92	0
10/31/96	23.35	−0.039056	1/2/97	25.69	−0.00891306
11/1/96	23.03	−0.0137993	1/3/97	25.59	−0.00390016
11/4/96	22.79	−0.0104759	1/6/97	26.37	0.0300254
11/5/96	22.64	−0.00660359	1/7/97	26.23	−0.00532321
11/6/96	22.69	0.00220605	1/8/97	26.62	0.014759
11/7/96	22.74	0.00220119	1/9/97	26.37	−0.00943581
11/8/96	23.59	0.0366974	1/10/97	26.09	−0.0106749
11/11/96	23.37	−0.00936974	1/13/97	25.19	−0.035105
11/12/96	23.35	−0.000856164	1/14/97	25.11	−0.00318092
11/13/96	24.12	0.0324444	1/15/97	25.95	0.0329054
11/14/96	24.41	0.0119515	1/16/97	25.52	−0.0167072
11/15/96	24.17	−0.00988069	1/17/97	25.41	−0.00431966
11/18/96	23.88	−0.0120709	1/20/97	25.23	−0.00710903
11/19/96	24.49	0.0252236	1/21/97	24.8	−0.0171901
11/20/96	23.76	−0.0302614	1/22/97	24.24	−0.0228395
11/21/96	23.84	0.00336135	1/23/97	24.18	−0.00247832
11/22/96	23.75	−0.00378231	1/24/97	24.05	−0.00539085
11/25/96	23.49	−0.0110077	1/27/97	23.94	−0.0045843

Table 10.5 *(cont.)*

Date	Price	Log Difference	Date	Price	Log Difference
1/28/97	23.9	−0.00167224	4/1/97	20.28	−0.0063898
1/29/97	24.47	0.0235694	4/2/97	19.47	−0.0407604
1/30/97	24.87	0.0162144	4/3/97	19.47	0
1/31/97	24.15	−0.0293779	4/4/97	19.12	−0.0181399
2/3/97	24.15	0	4/7/97	19.23	0.00573665
2/4/97	24.02	−0.00539756	4/8/97	19.35	0.00622086
2/5/97	23.91	−0.00459004	4/9/97	19.27	−0.00414294
2/6/97	23.1	−0.0344642	4/10/97	19.57	0.0154483
2/7/97	22.23	−0.0383899	4/11/97	19.53	−0.00204604
2/10/97	22.46	0.0102932	4/14/97	19.9	0.018768
2/11/97	22.42	−0.00178253	4/15/97	19.83	−0.00352379
2/12/97	21.86	−0.0252949	4/16/97	19.35	−0.0245035
2/13/97	22.02	0.00729265	4/17/97	19.42	0.00361104
2/14/97	22.41	0.0175562	4/18/97	19.91	0.0249187
2/17/97	22.41	0	4/21/97	20.38	0.0233319
2/18/97	22.52	0.00489652	4/22/97	19.6	−0.0390245
2/19/97	22.79	0.011918	4/23/97	19.73	0.00661075
2/20/97	21.98	−0.0361889	4/24/97	20.03	0.0150908
2/21/97	21.39	−0.0272094	4/25/97	19.99	−0.001999
2/24/97	20.71	−0.0323068	4/28/97	19.91	−0.00401003
2/25/97	21	0.0139058	4/29/97	20.44	0.0262716
2/26/97	21.11	0.00522442	4/30/97	20.21	−0.0113162
2/27/97	20.89	−0.0104763	5/1/97	19.91	−0.0149554
2/28/97	20.3	−0.0286497	5/2/97	19.6	−0.0156926
3/3/97	20.25	−0.00246609	5/5/97	19.63	0.00152944
3/4/97	20.66	0.0200447	5/6/97	19.66	0.00152711
3/5/97	20.49	−0.0082625	5/7/97	19.62	−0.00203666
3/6/97	20.94	0.0217242	5/8/97	20.34	0.0360399
3/7/97	21.28	0.0161065	5/9/97	20.43	0.00441502
3/10/97	20.49	−0.0378307	5/12/97	21.38	0.0454515
3/11/97	20.11	−0.0187198	5/13/97	21.37	−0.000467836
3/12/97	20.62	0.0250443	5/14/97	21.39	0.000935454
3/13/97	20.7	0.00387222	5/15/97	21.3	−0.00421645
3/14/97	21.29	0.0281038	5/16/97	22.12	0.0377751
3/17/97	20.92	−0.0175318	5/19/97	21.59	−0.0242519
3/18/97	22.06	0.0530604	5/20/97	21.19	−0.0187009
3/19/97	22.04	−0.00090703	5/21/97	21.86	0.0311291
3/20/97	22.32	0.0126242	5/22/97	21.86	0
3/21/97	21.51	−0.0369652	5/23/97	21.63	−0.0105772
3/24/97	21.06	−0.0211424	5/26/97	21.63	0
3/25/97	20.99	−0.00332937	5/27/97	20.79	−0.0396091
3/26/97	20.64	−0.0168152	5/28/97	20.79	0
3/27/97	20.7	0.00290276	5/29/97	20.97	0.00862074
3/28/97	20.7	0	5/30/97	20.88	−0.00430108
3/31/97	20.41	−0.0141087	6/2/97	21.12	0.0114287

Table 10.5 *(cont.)*

Date	Price	Log Difference	Date	Price	Log Difference
6/3/97	20.33	−0.0381228	8/5/97	20.81	0.00288739
6/4/97	20.12	−0.0103833	8/6/97	20.46	−0.0169619
6/5/97	19.66	−0.0231282	8/7/97	20.09	−0.0182496
6/6/97	18.79	−0.0452613	8/8/97	19.54	−0.0277585
6/9/97	18.68	−0.00587138	8/11/97	19.69	0.00764725
6/10/97	18.67	−0.000535475	8/12/97	19.99	0.0151213
6/11/97	18.53	−0.00752692	8/13/97	20.19	0.00995528
6/12/97	18.69	0.00859758	8/14/97	20.08	−0.00546314
6/13/97	18.83	0.00746272	8/15/97	20.07	−0.000498132
6/16/97	19.01	0.00951381	8/18/97	19.91	−0.00800404
6/17/97	19.23	0.0115064	8/19/97	20.12	0.0104922
6/18/97	18.79	−0.0231467	8/20/97	20.06	−0.00298656
6/19/97	18.67	−0.00640686	8/21/97	19.66	−0.0201417
6/20/97	18.55	−0.00644817	8/22/97	19.7	0.00203252
6/23/97	19.14	0.0313106	8/25/97	19.26	−0.0225882
6/24/97	19.03	−0.0057637	8/26/97	19.28	0.00103788
6/25/97	19.52	0.0254229	8/27/97	19.73	0.023072
6/26/97	19.09	−0.0222749	8/28/97	19.58	−0.00763168
6/27/97	19.46	0.0191964	8/29/97	19.61	0.001531
6/30/97	19.8	0.0173209	9/1/97	19.61	0
7/1/97	20.12	0.0160324	9/2/97	19.65	0.0020377
7/2/97	20.34	0.010875	9/3/97	19.61	−0.0020377
7/3/97	19.56	−0.0391027	9/4/97	19.4	−0.0107666
7/4/97	19.56	0	9/5/97	19.63	0.0117859
7/7/97	19.52	−0.00204708	9/8/97	19.45	−0.00921194
7/8/97	19.73	0.0107007	9/9/97	19.42	−0.00154361
7/9/97	19.46	−0.0137792	9/10/97	19.42	0
7/10/97	19.22	−0.0124097	9/11/97	19.37	−0.00257799
7/11/97	19.33	0.00570689	9/12/97	19.32	−0.00258465
7/14/97	18.99	−0.0177458	9/15/97	19.27	−0.00259135
7/15/97	19.67	0.0351821	9/16/97	19.61	0.0174902
7/16/97	19.65	−0.00101729	9/17/97	19.42	−0.00973618
7/17/97	19.99	0.0171548	9/18/97	19.38	−0.00206186
7/18/97	19.27	−0.0366827	9/19/97	19.35	−0.00154919
7/21/97	19.18	−0.00468141	9/22/97	19.6	0.0128371
7/22/97	19.08	−0.0052274	9/23/97	19.79	0.00964719
7/23/97	19.63	0.0284183	9/24/97	19.94	0.007551
7/24/97	19.77	0.00710663	9/25/97	20.39	0.0223168
7/25/97	19.89	0.00605146	9/26/97	20.87	0.0232681
7/28/97	19.81	−0.00403023	9/29/97	21.26	0.0185147
7/29/97	19.85	0.00201715	9/30/97	21.18	−0.00377003
7/30/97	20.3	0.0224169	10/1/97	21.05	−0.00615678
7/31/97	20.14	−0.007913	10/2/97	21.77	0.0336323
8/1/97	20.28	0.00692729	10/3/97	22.76	0.0444717
8/4/97	20.75	0.0229111	10/6/97	21.93	−0.037149

Table 10.5 *(cont.)*

Date	Price	Log Difference	Date	Price	Log Difference
10/7/97	21.96	0.00136705	10/29/97	20.71	0.0121449
10/8/97	22.18	0.00996837	10/30/97	21.22	0.0243275
10/9/97	22.12	−0.00270881	10/31/97	21.08	−0.00661941
10/10/97	22.1	−0.000904568	11/3/97	20.96	−0.00570886
10/13/97	21.32	−0.035932	11/4/97	20.7	−0.0124822
10/14/97	20.7	−0.0295119	11/5/97	20.31	−0.0190203
10/15/97	20.57	−0.0063	11/6/97	20.39	0.00393121
10/16/97	20.97	0.0192591	11/7/97	20.77	0.0184651
10/17/97	20.59	−0.0182873	11/10/97	20.4	−0.0179747
10/20/97	20.7	0.00532818	11/11/97	20.51	0.00537767
10/21/97	20.67	−0.00145033	11/12/97	20.49	−0.00097561
10/22/97	21.42	0.0356417	11/13/97	20.7	0.0101967
10/23/97	21.09	−0.0155261	11/14/97	21	0.0143887
10/24/97	20.97	−0.00570615	11/17/97	20.26	−0.0358739
10/27/97	21.07	0.00475738	11/18/97	20.04	−0.0109182
10/28/97	20.46	−0.0293785	11/19/97	19.8	−0.0120483

11. Autoregressive Models and Mean Reversion

11.1   The Autoregressive Model

Let $S_d(n)$ be the price of a security at the end of day n. If we also let

$$L(n) = \log(S_d(n)),$$

then the geometric Brownian motion model implies that

$$L(n) = a + L(n-1) + e(n), \qquad (11.1)$$

where $e(n)$, $n \geq 1$, is a sequence of independent and identically distributed normal random variables with mean 0 and variance σ^2/N (with $N = 252$ as the number of trading days in a year) and a is equal to μ/N. As before, μ is the mean (or drift) parameter of the geometric Brownian motion and σ is the associated volatility parameter.

Looking at Equation (11.1), it is natural to consider fitting a more general equation for $L(n)$; namely, the linear regression equation

$$L(n) = a + bL(n-1) + e(n), \qquad (11.2)$$

where b is another constant whose value would need to be estimated. That is, rather than arbitrarily taking $b = 1$, an improved model might be obtained by letting b's value be determined by data. Equation (11.2) is the classical linear regression model, and the technique for estimating a, b, and σ is well known. Because the linear regression model given by Equation (11.2) specifies the log price at time n in terms of the log price one time period earlier, it is called an *autoregressive model* of order 1.

The parameters a and b of the autoregressive model given by (11.2) are estimated from historical data in the following manner. Suppose $L(0), L(1), \ldots, L(r)$ are the logarithms of the end-of-day prices for r successive days. Then, when a and b are known, the predicted value of $L(i)$ based on prior log prices is $a + bL(i-1)$; hence, the usual approach to estimating a and b is to let them be the values that minimize the sum of squares of the prediction errors. That is, a and b are chosen to minimize

$$\sum_{i=1}^{r} (L(i) - a - bL(i-1))^2.$$

There are many standard statistical software packages that can be used to calculate the minimimizing values and also to estimate σ.

Remark. The model specified by Equation (11.2) is a risk-neutral model only when $a = (r - \sigma^2/2)/N$ and $b = 1$. That is, it is risk-neutral only when it reduces to the risk-neutral geometric Brownian motion model. Consequently, no arbitrage is possible when all investments are priced according to their expected present values when $a = (r - \sigma^2/2)/N$ and $b = 1$. However, an investor who believes that a and b have some other values can often make an investment that, although not yielding a sure win, can generate a return with a large expected value and a small variance when these latter quantities are computed according to the investor's estimated values of a and b.

11.2 Valuing Options by Their Expected Return

Assume that the end-of-day log prices follow Equation (11.2) and that the parameters a, b, σ have been determined, and consider an option whose exercise time is at the end of n trading days. In order to assess the expected value of this option's payoff, we must first determine the probability distribution of $L(n)$. To accomplish this, start by rewriting the Equation (11.2) as

$$L(i) = e(i) + a + bL(i-1).$$

Now, continually using the preceding equation – first with $i = n$, then with $i = n - 1$, and so on – yields

$$
\begin{aligned}
L(n) &= e(n) + a + bL(n-1) \\
&= e(n) + a + b[e(n-1) + a + bL(n-2)] \\
&= e(n) + be(n-1) + a + ab + b^2L(n-2) \\
&= e(n) + be(n-1) + a + ab + b^2[e(n-2) + a + bL(n-3)] \\
&= e(n) + be(n-1) + b^2e(n-2) \\
&\quad + a + ab + ab^2 + b^3L(n-3).
\end{aligned}
$$

Continuing on in this fashion shows that, for any $k < n$,

$$L(n) = \sum_{i=0}^{k} b^i e(n - i) + a \sum_{i=0}^{k} b^i + b^{k+1} L(n - k - 1).$$

Hence, with $k = n - 1$, the preceding equation yields

$$L(n) = \sum_{i=0}^{n-1} b^i e(n - i) + a \sum_{i=0}^{n-1} b^i + b^n L(0)$$

$$= \sum_{i=0}^{n-1} b^i e(n - i) + \frac{a(1 - b^n)}{1 - b} + b^n L(0). \qquad (11.3)$$

Note that $b^i e(n - i)$ is a normal random variable with mean 0 and variance $b^{2i} \sigma^2 / N$. Thus – using that the sum of independent normal random variables is also a normal random variable – we see that $\sum_{i=0}^{n-1} b^i e(n - i)$ is a normal random variable with mean

$$E\left[\sum_{i=0}^{n-1} b^i e(n - i) \right] = \sum_{i=0}^{n-1} b^i E[e(n - i)] = 0 \qquad (11.4)$$

and variance

$$\mathrm{Var}\left[\sum_{i=0}^{n-1} b^i e(n - i) \right] = \sum_{i=0}^{n-1} \mathrm{Var}[b^i e(n - i)]$$

$$= \frac{\sigma^2}{N} \sum_{i=0}^{n-1} b^{2i}$$

$$= \frac{\sigma^2 (1 - b^{2n})}{N(1 - b^2)}. \qquad (11.5)$$

Hence, from Equations (11.3), (11.4), and (11.5) we obtain that if the logarithm of the price at time 0 is $L(0) = g$, then $L(n)$ is a normal random variable with mean $m(n)$ and variance $v(n)$, where

$$m(n) = \frac{a(1 - b^n)}{1 - b} + b^n g \qquad (11.6)$$

and

$$v(n) = \frac{\sigma^2 (1 - b^{2n})}{N(1 - b^2)}. \qquad (11.7)$$

The present value of the payoff of a call option (whose strike price is
K and whose exercise time is at the end of n trading days) is

$$e^{-rn/N}(S_d(n) - K)^+ = e^{-rn/N}(e^{L(n)} - K)^+,$$

where r and N are (respectively) the interest rate and the number of trading days in a year. Using that $L(n)$ is normal with mean and variance as
given by Equations (11.6) and (11.7), it can be shown that the expected
value of this payoff is

$$E[e^{-rn/N}(e^{L(n)} - K)^+]$$
$$= e^{-rn/N}\big(e^{m(n)+v(n)/2}\Phi(\sqrt{v(n)} - h) - K\Phi(-h)\big), \quad (11.8)$$

where Φ is the standard normal distribution function and where

$$h = \frac{\log(K) - m(n)}{\sqrt{v(n)}}.$$

Example 11.2a   Assuming that an autoregressive model is appropriate
for the crude oil data from Chapter 10, the estimates of a, b, and $\sigma/\sqrt{N}$
obtained from a standard statistical package are

$$a = .0487, \quad b = .9838, \quad \sigma/\sqrt{N} = .01908.$$

That is, the estimated autoregressive equation is

$$L(n) = .0487 + .9838L(n - 1) + e(n),$$

where $e(n)$ is a normal random variable having mean 0 and standard
deviation .01908. Consequently, if the present price is 20, then the logarithm of the price at the end of another 50 trading days is a normal
random variable with mean

$$m(50) = \frac{.0487(1 - .9838^{50})}{1 - .9838} + \log(20)(.9838)^{50} = 3.0016$$

and variance

$$v(50) = (.0191)^2\frac{1 - (.9838)^{100}}{1 - (.9838)^2} = .0091.$$

Suppose now that the interest rate is 8% and that we want to determine
the expected present value of the payoff from an option to purchase the
security at the end of 50 trading days at a strike price $K = 21$. Because

$$h = \frac{\log(21) - 3.0016}{\sqrt{.0091}} = .4499,$$

it follows from Equation (11.8) that the present value of the expected payoff is

$$e^{-.08(50)/252}(20.2094\Phi(-.3545) - 21\Phi(-.4499)) = .4442.$$

That is, the expected present value payoff is 44.42 cents.

It is interesting to compare the preceding result with the geometric Brownian motion Black–Scholes option cost. Using the notation of Section 7.2, the data set of the crude oil prices results in the following estimate of the volatility parameter:

$$\sigma = .3032 \quad (\sigma/\sqrt{N} = .01910).$$

As this gives $\omega = -.1762$ and $\sigma\sqrt{t} = .1351$, the Black–Scholes cost is

$$C = 20\Phi(-.1762) - 21e^{-4/252}\Phi(-.3113) = .7911.$$

Thus the geometric Brownian motion risk-neutral cost valuation of 79 cents is quite a bit more than the expected present value payoff of 44 cents when the autoregressive model is assumed. The primary reason for this discrepancy is that the variance of the logarithm of the final price is .01824 under the risk-neutral geometric Brownian motion model but only .0091 under the autoregresssive model. (The means of the logarithms of the price at exercise time are roughly equal: 3.0025 under the risk-neutral geometric Brownian motion model and 3.0016 under the autoregressive model.)

For additional comparisons, a simulation study yielded that the expected present value of the option payoff under the model of Chapter 10 is 64 cents when the sample means are used as estimators of the mean drifts versus 81 cents when the risk-neutral means are used. □

11.3   Mean Reversion

Many traders believe that the prices of certain securities (often commodities) tend to revert to fixed values. That is, when the current price is less than this value, the price tends to increase; when it is greater, it tends to decrease. Although this phenomenon – called *mean reversion* – cannot be explained by a geometric Brownian motion model, it is a very simple consequence of the autoregressive model. For consider the model

$$L(n) = a + bL(n-1) + e(n),$$

which is equivalent to

$$S_d(n) = e^{a+e(n)}(S_d(n-1))^b.$$

Since

$$E[e^{a+e(n)}] = e^{a+\sigma^2/2N}$$

it follows that, if the price of the security at the end of day $n-1$ is s, then the expected price of the security at the end of the next day is

$$E[S_d(n)] = e^{a+\sigma^2/2N}s^b. \tag{11.9}$$

Now suppose that $0 < b < 1$, and let

$$s^* = \exp\left\{\frac{a+\sigma^2/2N}{1-b}\right\}.$$

We will show that if the present price is s then the expected price at the end of the next day is between s and s^*.

Toward this end, first suppose that $s < s^*$. That is,

$$s < \exp\left\{\frac{a+\sigma^2/2N}{1-b}\right\}, \tag{11.10}$$

which implies that

$$s^{1-b} < \exp\{a+\sigma^2/2N\}$$

or

$$s < \exp\{a+\sigma^2/2N\}s^b = E[S_d(n)]. \tag{11.11}$$

Moreover, Equation (11.10) also implies that

$$s^b < \exp\left\{\frac{b(a+\sigma^2/2N)}{1-b}\right\}$$

or

$$s^b < \exp\left\{\frac{a + \sigma^2/2N}{1 - b} - (a + \sigma^2/2N)\right\},$$

which is equivalent to

$$E[S_d(n)] = \exp\{a + \sigma^2/2N\}s^b < \exp\left\{\frac{a + \sigma^2/2N}{1 - b}\right\} = s^*. \quad (11.12)$$

Consequently, from (11.11) and (11.12) we see that, if $S_d(n - 1) = s < s^*$, then

$$s < E[S_d(n)] < s^*.$$

In a similar manner, it follows that if $S_d(n - 1) = s > s^*$ then

$$s^* < E[S_d(n)] < s.$$

Therefore, if $0 < b < 1$ then, for any current end-of-day price s, the mean price at the end of the next day is between s and s^*. In other words, there is a mean reversion to the price s^*.

Example 11.3a   For the data of Example 11.2a, the estimated regression equation is

$$L(n) = .0487 + .9838L(n - 1) + e(n),$$

where $e(n)$ is a normal random variable having mean 0 and standard deviation .0191. Since the estimated value of b is less than 1, this model predicts a mean price reversion to the value

$$s^* = \exp\left\{\frac{.0487 + (.0191)^2/2}{1 - .9838}\right\} = 20.44.$$

11.4   Exercises

Exercise 11.1   For the model

$$L(n) = 5 + .8L(n - 1) + e(n),$$

where $e(n)$ is a normal random variable with mean 0 and variance .2, find the probability that $L(n + 10) > L(n)$.

Exercise 11.2   Let $L(n)$ denote the logarithm of the price of a security at the end of day n, and suppose that

$$L(n) = 1.2 + .7L(n-1) + e(n),$$

where $e(n)$ is a normal random variable with mean 0 and variance .1. Find the expected present value payoff of a call option that expires in 60 trading days and has strike price 50 when the interest rate is 10% and the present price of the security is: (a) 48; (b) 50; (c) 52.

Exercise 11.3   Use a statistical package on the first 100 data values for heating oil (presented in Table 11.1, pp. 174–82) to fit an autoregressive model.

Exercise 11.4   To what value does the expected price of the security in Exercise 11.2 revert?

Exercise 11.5   For the model of Section 11.3, show that if $S_d(n-1) = s > s^*$ then

$$s^* < E[S_d(n)] < s.$$

Exercise 11.6   For the model of Section 11.3, show that if $S_d(n-1) = s^*$ then

$$E[S_d(n)] = s^*.$$

Table 11.1: *Nearest-Month Commodity Prices* (dollars)

Date	Unleaded Gas	Heating Oil	Date	Unleaded Gas	Heating Oil
03-Jan-95	52.75	49.94	07-Mar-95	56.78	46.36
04-Jan-95	53.43	49.64	08-Mar-95	55.83	45.25
05-Jan-95	54.51	49.96	09-Mar-95	54.35	45.14
06-Jan-95	53.77	49.52	10-Mar-95	52.47	45.25
09-Jan-95	53.9	48.33	13-Mar-95	53.81	45.61
10-Jan-95	53.66	47.38	14-Mar-95	52.79	44.34
11-Jan-95	54.54	47.98	15-Mar-95	54.04	45.14
12-Jan-95	54.92	47.85	16-Mar-95	54.93	45.37
13-Jan-95	55	46.68	17-Mar-95	55.37	46.07
16-Jan-95	56.88	47.35	20-Mar-95	56.15	45.85
17-Jan-95	57.8	48.67	21-Mar-95	56.15	45.65
18-Jan-95	59.48	49.08	22-Mar-95	55.9	47.02
19-Jan-95	58.12	48.28	23-Mar-95	57.53	46.56
20-Jan-95	57.4	48.14	24-Mar-95	57.82	46.32
23-Jan-95	56.38	47.82	27-Mar-95	58.6	47.46
24-Jan-95	57.6	47.87	28-Mar-95	58.73	47.46
25-Jan-95	57.25	47.47	29-Mar-95	59.99	47.08
26-Jan-95	57.44	47.27	30-Mar-95	60.68	47.19
27-Jan-95	56.07	47.27	31-Mar-95	59.47	47.06
30-Jan-95	56.21	47.42	03-Apr-95	57.44	47.47
31-Jan-95	57.76	46.86	04-Apr-95	58.6	47.96
01-Feb-95	56.77	47.8	05-Apr-95	60.48	48.01
02-Feb-95	55.95	48.55	06-Apr-95	61.68	49.21
03-Feb-95	57.35	49.44	07-Apr-95	61.29	49.5
06-Feb-95	57.3	49.2	10-Apr-95	61.22	49.28
07-Feb-95	56.99	49.13	11-Apr-95	61.59	50.15
08-Feb-95	56.1	47.98	12-Apr-95	61.37	49.54
09-Feb-95	55.84	47.65	13-Apr-95	60.44	48.79
10-Feb-95	55.64	48.28	14-Apr-95	60.44	48.79
13-Feb-95	55.56	47.29	17-Apr-95	62.03	50.01
14-Feb-95	56.16	47.5	18-Apr-95	63.69	50.19
15-Feb-95	56.22	46.89	19-Apr-95	63.15	50.15
16-Feb-95	57.91	46.92	20-Apr-95	63.22	50.28
17-Feb-95	58.76	47.72	21-Apr-95	63.2	50.64
20-Feb-95	58.76	47.72	24-Apr-95	62.21	50.02
21-Feb-95	59.11	47.62	25-Apr-95	62.91	50.78
22-Feb-95	59.84	47.89	26-Apr-95	63.81	50.45
23-Feb-95	58.36	47.44	27-Apr-95	64.96	51.26
24-Feb-95	58.76	47.75	28-Apr-95	65.33	51.19
27-Feb-95	58.97	47.19	01-May-95	64.15	51.09
28-Feb-95	57.58	46.9	02-May-95	63.65	50.95
01-Mar-95	56.74	46.44	03-May-95	62.55	50.25
02-Mar-95	55.59	46.52	04-May-95	63.59	51.27
03-Mar-95	55.94	47.41	05-May-95	63.99	51.34
06-Mar-95	56.21	46.66	08-May-95	64.21	51.15

Table 11.1 *(cont.)*

Date	Unleaded Gas	Heating Oil	Date	Unleaded Gas	Heating Oil
09-May-95	62.56	49.14	11-Jul-95	54.19	46.96
10-May-95	63.29	49.95	12-Jul-95	54.96	47.23
11-May-95	63.28	49.09	13-Jul-95	54.39	46.68
12-May-95	63.67	49.54	14-Jul-95	54.54	46.53
15-May-95	64.9	49.86	17-Jul-95	53.98	46.49
16-May-95	66.3	50.45	18-Jul-95	53.58	46.98
17-May-95	66.76	50.4	19-Jul-95	52.69	46.47
18-May-95	66.5	50.56	20-Jul-95	52.18	46.1
19-May-95	66.34	51.01	21-Jul-95	52.05	46.14
22-May-95	66.46	51.29	24-Jul-95	53.26	46.56
23-May-95	66.15	52.29	25-Jul-95	52.37	46.51
24-May-95	64.93	51.13	26-Jul-95	52.89	48.62
25-May-95	65.81	51.25	27-Jul-95	53.69	48.13
26-May-95	64.07	48.72	28-Jul-95	53.75	48
29-May-95	64.07	48.72	31-Jul-95	54.08	48.27
30-May-95	63.5	48.56	01-Aug-95	54.35	48.79
31-May-95	63	48.47	02-Aug-95	54.44	49.44
01-Jun-95	59.78	49.53	03-Aug-95	53.93	49.24
02-Jun-95	60.94	49.9	04-Aug-95	53.97	49.18
05-Jun-95	61.79	49.6	07-Aug-95	54.05	49.32
06-Jun-95	61.39	49.1	08-Aug-95	54.38	49.7
07-Jun-95	61.77	48.95	09-Aug-95	54.78	49.45
08-Jun-95	60.64	48.65	10-Aug-95	55.65	49.55
09-Jun-95	60.8	48.1	11-Aug-95	55.72	49.38
12-Jun-95	61.15	48.5	14-Aug-95	55.23	48.77
13-Jun-95	60.93	48.53	15-Aug-95	54.82	48.74
14-Jun-95	62	49.19	16-Aug-95	53.92	49.22
15-Jun-95	61.87	48.88	17-Aug-95	54.29	49.27
16-Jun-95	61.5	48.29	18-Aug-95	54.23	49.7
19-Jun-95	60.28	47	21-Aug-95	54.46	50.29
20-Jun-95	60.15	47.14	22-Aug-95	54.57	50.18
21-Jun-95	58.73	46.54	23-Aug-95	55.27	50.5
22-Jun-95	58.33	46.65	24-Aug-95	55.86	50.2
23-Jun-95	56.98	46.31	25-Aug-95	55.97	49.97
26-Jun-95	56.71	46.78	28-Aug-95	55.62	49.8
27-Jun-95	57.38	47.23	29-Aug-95	55.51	49.52
28-Jun-95	59.59	47.69	30-Aug-95	56.45	49.65
29-Jun-95	59.01	46.92	31-Aug-95	56.25	50.15
30-Jun-95	59.15	46.72	01-Sep-95	54.25	51.43
03-Jul-95	59.15	46.72	04-Sep-95	54.25	51.43
04-Jul-95	59.15	46.72	05-Sep-95	56.23	52.97
05-Jul-95	54.37	46.51	06-Sep-95	55.32	52.11
06-Jul-95	54.74	47.19	07-Sep-95	54.55	51.44
07-Jul-95	53.8	46.37	08-Sep-95	54.79	51.83
10-Jul-95	54.74	47.1	11-Sep-95	54.92	51.65

Table 11.1 *(cont.)*

Date	Unleaded Gas	Heating Oil	Date	Unleaded Gas	Heating Oil
12-Sep-95	55.74	51.95	14-Nov-95	50.43	51.56
13-Sep-95	55.34	51.25	15-Nov-95	51.24	51.71
14-Sep-95	56.81	51.8	16-Nov-95	51.55	52.22
15-Sep-95	56.63	51.53	17-Nov-95	52.79	52.96
18-Sep-95	57.73	51.65	20-Nov-95	52.9	52.73
19-Sep-95	57.23	51.37	21-Nov-95	53.12	52.28
20-Sep-95	56.39	49.3	22-Nov-95	54.12	52.54
21-Sep-95	54.87	48.67	23-Nov-95	54.12	52.54
22-Sep-95	53.49	48.09	24-Nov-95	54.12	52.54
25-Sep-95	54.01	48.85	27-Nov-95	55.45	53.42
26-Sep-95	53.79	48.23	28-Nov-95	56.24	52.95
27-Sep-95	54.55	49.02	29-Nov-95	57.45	52.2
28-Sep-95	56.05	49.5	30-Nov-95	57.36	51.62
29-Sep-95	57.67	48.65	01-Dec-95	53.02	52.67
02-Oct-95	52.78	49.26	04-Dec-95	53.56	54.03
03-Oct-95	51.93	49.28	05-Dec-95	54	54.22
04-Oct-95	50.74	48.85	06-Dec-95	53.89	54.75
05-Oct-95	48.89	47.97	07-Dec-95	54.06	55.28
06-Oct-95	49.15	48.21	08-Dec-95	54.65	56.59
09-Oct-95	50.24	48.74	11-Dec-95	54.69	56.75
10-Oct-95	50.33	48.67	12-Dec-95	55.58	56.81
11-Oct-95	50.48	48.8	13-Dec-95	57.55	57.69
12-Oct-95	49.86	48.46	14-Dec-95	57.86	57.3
13-Oct-95	50.29	48.92	15-Dec-95	59.59	57.99
16-Oct-95	50.7	48.85	18-Dec-95	59.93	59.11
17-Oct-95	50.33	48.82	19-Dec-95	59.26	59.23
18-Oct-95	49.88	48.42	20-Dec-95	57.75	59.9
19-Oct-95	49.36	48.15	21-Dec-95	56.91	60.01
20-Oct-95	49.7	48.58	22-Dec-95	57.59	60.09
23-Oct-95	49.81	48.94	25-Dec-95	57.59	60.09
24-Oct-95	49.87	49.36	26-Dec-95	58.69	60.5
25-Oct-95	49.69	49.58	27-Dec-95	60.26	62.33
26-Oct-95	50	50.44	28-Dec-95	59.28	60.32
27-Oct-95	50.06	50.34	29-Dec-95	58.6	58.63
30-Oct-95	50.74	50.59	01-Jan-96	58.6	58.63
31-Oct-95	50.83	50.4	02-Jan-96	59.09	59.93
01-Nov-95	50.55	50.95	03-Jan-96	58.74	59.44
02-Nov-95	51.72	52.04	04-Jan-96	59.44	59.28
03-Nov-95	51.51	51.72	05-Jan-96	60.48	60.64
06-Nov-95	51.03	51.15	08-Jan-96	60.48	60.64
07-Nov-95	51.14	50.99	09-Jan-96	58.65	60.43
08-Nov-95	51.42	51.45	10-Jan-96	58.19	59.59
09-Nov-95	51.06	51.62	11-Jan-96	54.44	56.16
10-Nov-95	50.7	51.63	12-Jan-96	53.1	53.57
13-Nov-95	50.3	51.57	15-Jan-96	53.9	53.3

Table 11.1 *(cont.)*

Date	Unleaded Gas	Heating Oil	Date	Unleaded Gas	Heating Oil
16-Jan-96	53.33	52.43	19-Mar-96	65.15	62.26
17-Jan-96	54.98	53.13	20-Mar-96	64.38	63.12
18-Jan-96	55.21	54.37	21-Mar-96	64.03	61.33
19-Jan-96	55.41	54.22	22-Mar-96	65.49	62.65
22-Jan-96	54.88	53.67	25-Mar-96	67	63.2
23-Jan-96	53.66	52.95	26-Mar-96	66.25	64.88
24-Jan-96	54.2	52.72	27-Mar-96	65.72	65.93
25-Jan-96	52.67	50.51	28-Mar-96	64.44	63.54
26-Jan-96	52.97	50.93	29-Mar-96	64.94	62.76
29-Jan-96	52.46	51.13	01-Apr-96	66	57.98
30-Jan-96	53.37	52.28	02-Apr-96	68.11	59.72
31-Jan-96	54.1	53.51	03-Apr-96	67.69	58.22
01-Feb-96	53.14	52.41	04-Apr-96	68.76	59.57
02-Feb-96	53.74	53.26	05-Apr-96	68.76	59.57
05-Feb-96	52.06	51.6	08-Apr-96	69.86	60.19
06-Feb-96	52.38	51.64	09-Apr-96	70.52	60.64
07-Feb-96	52.23	52.46	10-Apr-96	72.99	62.51
08-Feb-96	52.44	53.14	11-Apr-96	74.3	64.02
09-Feb-96	52.91	53.62	12-Apr-96	72.17	62.02
12-Feb-96	53	53.69	15-Apr-96	71.71	62.62
13-Feb-96	55.11	56.74	16-Apr-96	69.45	59.54
14-Feb-96	55.2	58.21	17-Apr-96	68.12	58.09
15-Feb-96	55.44	57	18-Apr-96	66.4	55.4
16-Feb-96	55.77	56.87	19-Apr-96	67.49	55.72
19-Feb-96	55.77	56.87	22-Apr-96	70.19	55.06
20-Feb-96	57.71	56.39	23-Apr-96	73.18	57.3
21-Feb-96	59.45	58.84	24-Apr-96	74.1	58.2
22-Feb-96	60.04	60.53	25-Apr-96	75.61	58.76
23-Feb-96	58.73	60.66	26-Apr-96	76.81	59.27
26-Feb-96	59.76	62.85	29-Apr-96	77.01	62.28
27-Feb-96	60.31	64.28	30-Apr-96	72.39	61.82
28-Feb-96	59.46	59.68	01-May-96	67.42	54.16
29-Feb-96	59.35	61.81	02-May-96	68.4	53.94
01-Mar-96	59.75	53.42	03-May-96	69.92	54.74
04-Mar-96	58.73	52.15	06-May-96	68.85	54.56
05-Mar-96	59.09	53	07-May-96	68.81	54.79
06-Mar-96	59.75	54.22	08-May-96	68.37	54.87
07-Mar-96	59.18	53.78	09-May-96	67.23	54.56
08-Mar-96	58.75	53.44	10-May-96	68.48	54.95
11-Mar-96	59.32	55.15	13-May-96	69.11	56.19
12-Mar-96	60.56	54.83	14-May-96	68.43	55.32
13-Mar-96	61.61	54.59	15-May-96	67.2	54.81
14-Mar-96	62.45	55.07	16-May-96	64.2	53
15-Mar-96	62.92	57.87	17-May-96	63.03	52.94
18-Mar-96	64.31	60.28	20-May-96	66.04	55.24

Table 11.1 *(cont.)*

Date	Unleaded Gas	Heating Oil	Date	Unleaded Gas	Heating Oil
21-May-96	64.95	54.06	23-Jul-96	63.08	55.94
22-May-96	64.3	54.99	24-Jul-96	61.87	55.95
23-May-96	64.25	54.39	25-Jul-96	61.66	56.25
24-May-96	64.72	54.46	26-Jul-96	60.16	55.04
27-May-96	64.72	54.46	29-Jul-96	60.52	55.19
28-May-96	63.15	54.18	30-Jul-96	61.23	55.65
29-May-96	62.36	54.06	31-Jul-96	61.8	57.08
30-May-96	59.88	52.09	01-Aug-96	61.38	57.53
31-May-96	59.12	50.85	02-Aug-96	62.12	58.71
03-Jun-96	58.99	51.25	05-Aug-96	61.31	58.29
04-Jun-96	60.69	51.52	06-Aug-96	61.23	57.43
05-Jun-96	59.39	50.85	07-Aug-96	62	58.22
06-Jun-96	60.22	51.04	08-Aug-96	62.27	58.79
07-Jun-96	60.91	51.78	09-Aug-96	61.87	58.49
10-Jun-96	61.4	51.4	12-Aug-96	62.89	59.56
11-Jun-96	60.8	50.79	13-Aug-96	63.09	60.01
12-Jun-96	59.68	50.88	14-Aug-96	62.49	60.41
13-Jun-96	58.89	50.95	15-Aug-96	61.96	59.68
14-Jun-96	59.5	51.55	16-Aug-96	63.38	61.63
17-Jun-96	61.21	53.34	19-Aug-96	65.27	62.58
18-Jun-96	60.24	52.5	20-Aug-96	64.01	61.67
19-Jun-96	57.96	51.12	21-Aug-96	63.12	60.98
20-Jun-96	58.68	51.53	22-Aug-96	63.88	62.48
21-Jun-96	58.74	51.36	23-Aug-96	63.22	61.99
24-Jun-96	58.23	51.3	26-Aug-96	61.62	61.03
25-Jun-96	57.46	51.17	27-Aug-96	61.21	61.13
26-Jun-96	58.36	52.34	28-Aug-96	62.33	62.04
27-Jun-96	59.36	53.64	29-Aug-96	63.72	63.67
28-Jun-96	60.03	53.95	30-Aug-96	62.82	62.82
01-Jul-96	61.51	55.14	02-Sep-96	62.82	62.82
02-Jul-96	60.89	54.28	03-Sep-96	62.96	65.07
03-Jul-96	62.47	54.71	04-Sep-96	62.96	64.21
04-Jul-96	62.47	54.71	05-Sep-96	64.41	65.03
05-Jul-96	62.47	54.71	06-Sep-96	65.27	66.4
08-Jul-96	61.68	54.89	09-Sep-96	64.09	65.95
09-Jul-96	61.81	55.26	10-Sep-96	64.85	66.67
10-Jul-96	63.11	55.59	11-Sep-96	65.91	68.19
11-Jul-96	64.59	56.7	12-Sep-96	65.91	69.17
12-Jul-96	64	56.62	13-Sep-96	64.6	67.94
15-Jul-96	65.56	57.72	16-Sep-96	62.87	65.29
16-Jul-96	65.13	57.18	17-Sep-96	62.74	65.59
17-Jul-96	63.89	56.32	18-Sep-96	63.06	67.87
18-Jul-96	63.87	56.74	19-Sep-96	61.32	66.77
19-Jul-96	62.41	56.02	20-Sep-96	61.09	67.42
22-Jul-96	62.76	55.85	23-Sep-96	60.07	67.48

Table 11.1 *(cont.)*

Date	Unleaded Gas	Heating Oil	Date	Unleaded Gas	Heating Oil
24-Sep-96	62.83	69.69	26-Nov-96	69.01	71.24
25-Sep-96	63.1	71.77	27-Nov-96	69.35	71.97
26-Sep-96	62.99	70.9	28-Nov-96	69.35	71.97
27-Sep-96	64.6	71.49	29-Nov-96	69.35	71.97
30-Sep-96	62.71	71.51	02-Dec-96	68.12	73.57
01-Oct-96	62.82	70.76	03-Dec-96	69.13	74.22
02-Oct-96	62.42	71.98	04-Dec-96	68.24	73.57
03-Oct-96	63.68	74.69	05-Dec-96	69.68	75.11
04-Oct-96	63.63	74.43	06-Dec-96	69.8	74.66
07-Oct-96	66.34	76.49	09-Dec-96	68.88	72.13
08-Oct-96	66.5	76.19	10-Dec-96	66.86	69.62
09-Oct-96	65.59	73.97	11-Dec-96	63.56	66.82
10-Oct-96	63.52	70.92	12-Dec-96	64.72	68.67
11-Oct-96	65.52	71.43	13-Dec-96	67.04	71.71
14-Oct-96	67.7	74.07	16-Dec-96	69.52	74.82
15-Oct-96	67.08	73.07	17-Dec-96	69.77	73.54
16-Oct-96	65.45	71.56	18-Dec-96	71.17	74.18
17-Oct-96	66.53	72.29	19-Dec-96	71.22	73.78
18-Oct-96	67.94	74.06	20-Dec-96	70.19	72.97
21-Oct-96	67.92	73.63	23-Dec-96	68.9	71.08
22-Oct-96	69.12	73.45	24-Dec-96	69.56	71.4
23-Oct-96	68.16	70.96	25-Dec-96	69.56	71.4
24-Oct-96	69.22	70.49	26-Dec-96	69.51	70.06
25-Oct-96	70.1	71.72	27-Dec-96	69.74	70.55
28-Oct-96	70.3	71.46	30-Dec-96	69.61	70.57
29-Oct-96	69.1	69.83	31-Dec-96	70.67	72.84
30-Oct-96	70	68.46	01-Jan-97	70.67	72.84
31-Oct-96	66.56	66.34	02-Jan-97	71.1	72.11
01-Nov-96	64.7	66.6	03-Jan-97	70.7	71.29
04-Nov-96	65	65.95	06-Jan-97	72.52	73.64
05-Nov-96	64.61	65.42	07-Jan-97	72.1	72.49
06-Nov-96	63.63	66.45	08-Jan-97	72.19	73.43
07-Nov-96	63.8	66.89	09-Jan-97	70.48	73.05
08-Nov-96	65.27	68.93	10-Jan-97	70.36	72.15
11-Nov-96	65.02	68.35	13-Jan-97	68.09	69.7
12-Nov-96	65.77	68.25	14-Jan-97	67.04	69.42
13-Nov-96	68.34	71.2	15-Jan-97	68.85	71.42
14-Nov-96	68.92	73.4	16-Jan-97	68.69	69.92
15-Nov-96	66.92	72.61	17-Jan-97	68.09	68.44
18-Nov-96	65.77	71.85	20-Jan-97	67.23	66.94
19-Nov-96	67.39	73.68	21-Jan-97	67.44	66.03
20-Nov-96	65.39	72.09	22-Jan-97	68.22	66.89
21-Nov-96	67.04	73.85	23-Jan-97	68.42	66.35
22-Nov-96	67.8	72.79	24-Jan-97	67.75	66.77
25-Nov-96	67.99	72.23	27-Jan-97	67.62	67.29

Table 11.1 *(cont.)*

Date	Unleaded Gas	Heating Oil	Date	Unleaded Gas	Heating Oil
28-Jan-97	67.04	66.83	01-Apr-97	62.67	53.95
29-Jan-97	68.23	68.84	02-Apr-97	60.61	52.52
30-Jan-97	69.82	70.34	03-Apr-97	60.9	53.26
31-Jan-97	68.47	68.65	04-Apr-97	60.48	53.14
03-Feb-97	68.35	65.28	07-Apr-97	60.72	53.13
04-Feb-97	68.31	64.18	08-Apr-97	61.17	52.89
05-Feb-97	67.54	63.32	09-Apr-97	60.7	53.11
06-Feb-97	65.3	61.45	10-Apr-97	61.07	54.86
07-Feb-97	63.06	60.53	11-Apr-97	60.88	53.87
10-Feb-97	63.53	61.76	14-Apr-97	61.96	54.67
11-Feb-97	63.96	61.86	15-Apr-97	61.9	54.85
12-Feb-97	62.89	60.85	16-Apr-97	60.38	53.48
13-Feb-97	63.18	59.92	17-Apr-97	60.7	54
14-Feb-97	64.25	60.81	18-Apr-97	61.49	54.68
17-Feb-97	64.25	60.81	21-Apr-97	62.8	55.48
18-Feb-97	64.16	59.42	22-Apr-97	61.77	54.83
19-Feb-97	64.68	59.59	23-Apr-97	61.74	55.65
20-Feb-97	62.78	58.04	24-Apr-97	62.84	55.89
21-Feb-97	61.82	57.85	25-Apr-97	62.5	55.9
24-Feb-97	60.24	55.47	28-Apr-97	62.34	56.53
25-Feb-97	62.23	56.82	29-Apr-97	63.36	58.91
26-Feb-97	62.26	56.68	30-Apr-97	63.91	58.07
27-Feb-97	62.67	56.03	01-May-97	62.63	54.33
28-Feb-97	61.65	54.76	02-May-97	60.52	53.02
03-Mar-97	61.77	53.18	05-May-97	60.54	53.05
04-Mar-97	62.89	53.34	06-May-97	60.31	53.53
05-Mar-97	63.33	52.54	07-May-97	60.92	53.08
06-Mar-97	64.48	53.43	08-May-97	62.5	54.38
07-Mar-97	65.67	54.08	09-May-97	62.89	54.52
10-Mar-97	64.36	53.08	12-May-97	64.47	56.65
11-Mar-97	63.86	52.83	13-May-97	64.76	56.48
12-Mar-97	64.63	54.08	14-May-97	64.38	56.42
13-Mar-97	64.23	54.22	15-May-97	64.04	56.48
14-Mar-97	65.77	55.33	16-May-97	65.87	58.47
17-Mar-97	65.26	54.3	19-May-97	65.21	57.92
18-Mar-97	67.48	56.18	20-May-97	65.39	57.64
19-Mar-97	67.96	56.29	21-May-97	66.53	57.55
20-Mar-97	67.58	55.94	22-May-97	66.93	57.8
21-Mar-97	67.64	55.98	23-May-97	66.92	57.52
24-Mar-97	66.51	55.73	26-May-97	66.92	57.52
25-Mar-97	66.52	56.83	27-May-97	65.38	55.27
26-Mar-97	64.82	55.43	28-May-97	65.8	55.39
27-Mar-97	64.63	56.07	29-May-97	65.15	56
28-Mar-97	64.63	56.07	30-May-97	63.68	56.49
31-Mar-97	63.68	56.72	02-Jun-97	63.68	56.32

Table 11.1 *(cont.)*

Date	Unleaded Gas	Heating Oil	Date	Unleaded Gas	Heating Oil
03-Jun-97	61.44	54.62	05-Aug-97	67.1	58.32
04-Jun-97	60.42	54.16	06-Aug-97	66.06	56.98
05-Jun-97	59.82	53.32	07-Aug-97	64.33	55.3
06-Jun-97	57.13	51.52	08-Aug-97	61.99	54.29
09-Jun-97	56.2	51.5	11-Aug-97	61.47	54.36
10-Jun-97	56.4	51.65	12-Aug-97	63.71	55.1
11-Jun-97	56.54	51.52	13-Aug-97	66.08	56.04
12-Jun-97	57.08	51.62	14-Aug-97	66.33	55.87
13-Jun-97	57.4	51.64	15-Aug-97	66.81	55.25
16-Jun-97	58.03	51.94	18-Aug-97	65.44	55.09
17-Jun-97	58.48	52.45	19-Aug-97	67.58	55.71
18-Jun-97	56.78	51.44	20-Aug-97	69.64	55.1
19-Jun-97	56.09	51.45	21-Aug-97	67.15	53.48
20-Jun-97	55.48	51.33	22-Aug-97	67.48	53.41
23-Jun-97	55.64	51.92	25-Aug-97	64.5	52.2
24-Jun-97	55.68	51.57	26-Aug-97	63.81	52.09
25-Jun-97	56.97	52.99	27-Aug-97	66.4	53.26
26-Jun-97	56.79	52.02	28-Aug-97	67.51	52.51
27-Jun-97	57.91	53.33	29-Aug-97	68.82	51.85
30-Jun-97	58.12	53.7	01-Sep-97	68.82	51.85
01-Jul-97	58.78	54.84	02-Sep-97	62.79	53.4
02-Jul-97	59.29	54.92	03-Sep-97	62.55	53.35
03-Jul-97	57.92	52.76	04-Sep-97	59.92	52.54
04-Jul-97	57.92	52.76	05-Sep-97	60.12	53.78
07-Jul-97	57.94	52.78	08-Sep-97	59.32	53.14
08-Jul-97	58.92	53	09-Sep-97	59.49	52.83
09-Jul-97	58.23	52.65	10-Sep-97	58.33	51.57
10-Jul-97	58.6	52.11	11-Sep-97	58.78	52.05
11-Jul-97	59.26	52.35	12-Sep-97	58.77	52.58
14-Jul-97	58.35	51.67	15-Sep-97	58.22	52.52
15-Jul-97	59.94	52.95	16-Sep-97	59.04	53.85
16-Jul-97	60.46	52.68	17-Sep-97	58.45	53.35
17-Jul-97	61.89	53.89	18-Sep-97	57.25	53.44
18-Jul-97	60.05	52.22	19-Sep-97	57.48	53.45
21-Jul-97	60.04	52.35	22-Sep-97	58.58	54.73
22-Jul-97	60.02	52.7	23-Sep-97	58.36	54.64
23-Jul-97	61.12	53.28	24-Sep-97	58.37	55.24
24-Jul-97	62.21	53.39	25-Sep-97	59.25	56.51
25-Jul-97	64.03	53.99	26-Sep-97	61.34	57.92
28-Jul-97	64.84	54.14	29-Sep-97	63.13	59.25
29-Jul-97	66.47	54.31	30-Sep-97	62.63	58.77
30-Jul-97	69.9	55.78	01-Oct-97	59.9	58.19
31-Jul-97	67.84	55.61	02-Oct-97	61.43	59.8
01-Aug-97	65.07	56.56	03-Oct-97	62.99	62.01
04-Aug-97	66.74	58.44	06-Oct-97	61.3	59.69

Table 11.1 *(cont.)*

Date	Unleaded Gas	Heating Oil	Date	Unleaded Gas	Heating Oil
07-Oct-97	60.91	59.6	07-Nov-97	59.95	57.99
08-Oct-97	61.5	60.16	10-Nov-97	59.18˙	57.28
09-Oct-97	61.18	60.08	11-Nov-97	59.06	57.82
10-Oct-97	61.24	59.95	12-Nov-97	58.62	57.92
13-Oct-97	59.83	58.27	13-Nov-97	59.55	58.62
14-Oct-97	58.88	57.01	14-Nov-97	60.99	59.54
15-Oct-97	58.2	56.94	17-Nov-97	59.44	57.85
16-Oct-97	59.68	58.01	18-Nov-97	58.65	57.61
17-Oct-97	59.31	57.4	19-Nov-97	58.65	56.67
20-Oct-97	59.66	57.82	20-Nov-97	57.22	55.45
21-Oct-97	59.08	57.64	21-Nov-97	57.74	55.48
22-Oct-97	60.79	58.77	24-Nov-97	58.69	55.6
23-Oct-97	60.26	58.09	25-Nov-97	59.08	55.49
24-Oct-97	59.6	57.03	26-Nov-97	57.31	53.1
27-Oct-97	59.95	57.74	27-Nov-97	57.31	53.1
28-Oct-97	58.88	56.52	28-Nov-97	57.31	53.1
29-Oct-97	60.09	57.19	01-Dec-97	56.25	52.71
30-Oct-97	60.67	58.12	02-Dec-97	56.43	53.25
31-Oct-97	60.22	57.77	03-Dec-97	56.55	53.5
03-Nov-97	59.8	58.78	04-Dec-97	56.34	53.35
04-Nov-97	58.96	58.11	05-Dec-97	56.59	53.38
05-Nov-97	58.2	57.18	08-Dec-97	56.96	53.52
06-Nov-97	59.26	57.43			

Index